"An Honorable Place in American Air Power"

Civil Air Patrol Coastal Patrol Operations, 1942–1943

Frank A. Blazich Jr., PhD
Colonel, CAP

Air University Press
Maxwell Air Force Base, Alabama

Air University Press

Director
Maj Richard Harrison

Managing Editor
Dr. Christopher Rein

Design and Production Managing Editor
Luetwinder Eaves

Project Editor
Donna Budjenska

Illustrator
Daniel Armstrong

Print Specialists
Megan Hoehn
Nedra Looney

Air University Press
600 Chennault Circle, Building 1405
Maxwell AFB, AL 36112-6010

https://www.airuniversity.af.edu/AUPress/

Facebook:
https://www.facebook.com/AirUnivPress
and
Twitter: https://twitter.com/aupress

AIR UNIVERSITY PRESS

Library of Congress Cataloging-in-Publication Data

Names: Blazich, Frank A., Jr., author. | Air University (U.S.). Press, issuing body.
Title: "An honorable place in American air power" : Civil Air Patrol coastal patrol operations, 1942-1943 / Frank A. Blazich, Jr.
Other titles: Civil Air Patrol coastal patrol operations, 1942-1943
Description: [Maxwell Air Force Base, Ala.] : [Air University Press], [2020] | Includes bibliographical references and index. | Summary: "Military historian and Civil Air Patrol (CAP) member Frank A. Blazich Jr. collects oral and written histories of the CAP's short-lived—but influential—coastal air patrol operations of World War II and expands it in a scholarly monograph that cements the legacy of this vital civil-military cooperative effort"— Provided by publisher.
Identifiers: LCCN 2020019712 (print) | LCCN 2020019713 (ebook) | ISBN 9781585663057 (paperback) | ISBN 9781585663057 (Adobe PDF) . | Air defenses—United States—History. | Civil defense—United States—History. | Anti-submarine warfare—United States—History. | World War, 1939-1945—United States.
Classification: LCC UG733 .B53 2020 (print) | LCC UG733 (ebook) | DDC 358.41450973—dc23 | SUDOC D 301.26/6:C 49/2
LC record available at https://lccn.loc.gov/2020019712
LC ebook record available at https://lccn.loc.gov/2020019713

Published by Air University Press in December 2020

Disclaimer

Opinions, conclusions, and recommendations expressed or implied within are solely those of the authors and do not necessarily represent the official policy or position of the organizations with which they are associated or the views of the Air University Press, Air University, United States Air Force, Department of Defense, or any other US government agency. This publication is cleared for public release and unlimited distribution.

This book and other Air University Press publications are available electronically at the AU Press website: https://www.airuniversity.af.edu/AUPress.

Contents

	List of Illustrations	*v*
	Foreword	*vii*
	About the Author	*ix*
	Preface	*xi*
	Acknowledgments	*xiii*
	Abbreviations	*xv*
1	Introduction	1
2	Origins of the Civil Air Patrol	11
3	*Paukenschlag* and the CAP Experiment	39
4	Learning and Expansion	67
5	From Maine to Mexico	99
6	Challenges and Transitions	129
7	Past Reflections and Future Possibilities	161
	Appendix A: Activation of Civil Air Patrol (CAP) Coastal Patrols	185
	Appendix B: CAP Coastal Patrol Personnel Killed on Active Duty	187
	Appendix C: CAP Coastal Patrol Base Commanders March 1942–August 1943	189
	Appendix D: CAP Coastal Patrol Flight Hours October 1942–August 1943	191
	Appendix E: Aircraft on Active CAP Coastal Duty as of 28 April 1943	193
	Appendix F: "Definitely Damaged or Destroyed": Reexamining CAP's Wartime Claims	197

CONTENTS

Appendix G: Proposed Plan for Organization of Civil Air Defense Resources   205

Bibliography   223

Index   235

Illustrations

Figures

1	Gill Robb Wilson	13
2	Milton Knight	15
3	Maj Gen John F. Curry	27
4	Civil Air Patrol organizational chart as of December 1941	30
5	Letter from William D. Mason to Earle L. Johnson of 4 February 1942	48
6	Sectional chart used for navigation by 1st Lt Henry E. Phipps	54
7	First Task Force, Atlantic City, New Jersey, 13 March 1942,	55
8	Flight line at the First Task Force	56
9	Radio testing and maintenance	71
10	Two-ship formation on patrol	74
11	Col Earle L. Johnson	76
12	Fourth Task Force, Parksley, Virginia	77
13	Tanker *Eclipse* torpedoed on 4 May 1942	86
14	Oil slick and debris from the merchant steamer *Ohioan*	87
15	Stinson Voyager 10A armed	88
16	Ninth Task Force, Grand Isle, Louisiana	101
17	Tech Sgt Addis H. McDonald with oil-soaked life jacket	102
18	Capt Warren E. Moody	107

ILLUSTRATIONS

19	President Roosevelt awarding Air Medal to CAP personnel	112
20	Map of coastal patrol base locations, September 1942.	113
21	Cpl Louise T. Story at teletypewriter	117
22	Mechanics working on engines	131
23	Recovered wreckage of crashed CAP coastal patrol aircraft	132
24	Fabric replacement on aircraft	133
25	Aircrew at mission briefing	137
26	Stinson Voyager 10A in three-tone camouflage	141
27	Recovery of *Panam* survivors	147
28	Lowering of flag at Parksley, Virgina	150
29	CAP Cessna intercepted by F-16	171
30	WaldoAir camera systems installed on a CAP Cessna 182-T	174
31	CAP Cessnas and MQ-9 Reaper of NY National Guard	176
32	CAP Cessna 172 P of Connecticut Wing in Operation Bird Dog	178

Foreword

This volume relates the proud history of an important organization, the Civil Air Patrol (CAP), and its efforts to use civilian-owned aircraft and volunteer pilots to help combat the U-boat menace that threatened America's shores in World War II. Though that story is not well known, it has been chronicled previously, but not with Frank Blazich's attention to detail and important corrections to CAP's effectiveness during the war. CAP's substantial accomplishments will be familiar to members of that fine organization that does much to inculcate a spirit of "airmindedness," as Billy Mitchell called it, among today's youth and motivate them towards careers in both civil and military aviation. But Blazich has gone far beyond a statistical recounting of sorties and hours flown, of the long tedious hours of patrol, and the sheer terror of engine failure far out at sea or an airfield closed in by weather as the fuel gauge approaches empty. Instead, he uses the successful mobilization of civilian "experts" (and they certainly knew more about aviation than many members of the general population) to come to their nation's aid in a time of crisis. By doing so, he reminds future commanders and planners to consider the use of civil resources and highlights issues that are likely to emerge in mobilizing these important assets, from the legal status of noncombatants to the importance of logistical support and sustainment.

The potential uses of civilian aviation experts in future crises are limited only by the imagination. The Civil Air Patrol, as currently organized and equipped, could easily provide reconnaissance and light logistical support in the event of war. CAP aircraft, manned and unmanned, fixed-wing or rotary, could be used to in real time verify or dispel "deep fakes" of events, helping combat the information warfare our adversaries are becoming so adept at. These same platforms could provide vital reconnaissance during natural disasters, from delivering life-saving medical supplies to relieving much more expensive and already heavily tasked military assets from the burdens of searching for survivors or compiling imagery of damage. As climate change threatens our globe, CAP orbits equipped with thermal sensors could help monitor forests for wildfires during critical periods so they can be extinguished before they become life-threatening infernos. And aircraft could assist in securing the nation's porous borders, whether on land or at sea. Similarly, civilian cyber specialists could

FOREWORD

lend their expertise in times of cyberattack to help defend the nation's vital economic and communications infrastructure.

Thus Frank Blazich's work is not only an excellent history of events over 75 years ago, but it is also a blueprint for leveraging all aspects of our national power in times of crisis. It will certainly appeal to the membership of the Civil Air Patrol, who will find inspiration from the sacrifices of a previous generation of Airmen, 68 of whom gave their lives in their nation's defense. But it will also help inform current and future commanders, planners, and civilian leaders on the capabilities of this remarkable organization and provide suggestions for the incorporation of civil assets to support future military and disaster relief operations, whether in air, space, or cyberspace. Accordingly, Air University Press is proud to publish "*An Honorable Place in American Air Power.*"

CHRISTOPHER M. REIN
Managing Editor, Air University Press

About the Author

Dr. Frank A. Blazich Jr. (BA, University of North Carolina at Chapel Hill; MA, North Carolina State University; PhD, The Ohio State University) is a curator of modern military history at the Smithsonian Institution's National Museum of American History. A veteran of the US Air Force, he had his first edited book, *Bataan Survivor: A POW's Account of Japanese Captivity in World War II*, published by the University of Missouri Press in February 2017. He has published widely and delivered public talks on numerous topics relating to modern American military history. His work has appeared in various publications, including the *North Carolina Historical Review*, *The Northern Mariner*, *Marine Corps History*, the *Naval War College Review*, *Air Power History*, *Army History*, *War on the Rocks*, and the *Washington Post*. In his free time, he is a colonel in the Civil Air Patrol and director of the Col Louisa S. Morse Center for Civil Air Patrol History. He previously served as the corporation's National Historian from April 2013 to March 2018, during which time he served as principal historical advisor for the Civil Air Patrol Congressional Gold Medal effort.

Preface

On the eve of American entry into World War II, the Office of Civilian Defense (OCD) established the Civil Air Patrol (CAP), an organization of the nation's private pilots and aviation personnel for national defense purposes. Beginning in February 1942 with tentative Navy Department approval, the Army Air Forces agreed to an experimental use of CAP aircraft and personnel for antisubmarine patrols along the Atlantic Coast. This use of civilian pilots and aircraft developed out of an urgent necessity to stem the tide of German submarine operations inflicting heavy losses on coastal shipping. For the Army, the CAP coastal patrol was essentially a subexperiment for a larger experiment to see if civilian aviation could be semimilitarized for national defense purposes.

The operational success of CAP's coastal patrol effort convinced Army leadership that the CAP could serve in a wider capacity. The coastal patrol effort received ordnance and military uniforms and expanded to 21 bases flying continuous daytime patrols from Maine to the Texas–Mexico border. The coastal patrol effort spawned a similar Southern Liaison Patrol patrolling the American border with Mexico. Collectively, CAP's operations with the Army resulted in the transfer of CAP from OCD to the War Department in late April 1943.

Drawing extensively on unpublished, previously unavailable archival material, this policy-based study of CAP's coastal patrol examines the origins, evolution, and concluding operations of this civilian effort. Through the historical record, conclusions are drawn from CAP's coastal patrol operation to provide a doctrinal basis for the discussion of future uses of auxiliary airmen for domestic military purposes in time of war.

Acknowledgments

I wish to thank my wife, Nicole M. Warburton, for her support and love in seeing this project through to publication. To my mother and late father, thank you for inculcating a love of history and the North Carolina Outer Banks, whence my research curiosity in the Civil Air Patrol first took flight. Numerous thanks are owed to my professional colleagues in CAP and academia for providing intellectual, archival, editorial, and moral support for this research endeavor. Among the many individuals who have taken time to provide me feedback or support, special thanks are extended to Lt Col Winton Adcock; Capt Jessica Allen; Col Jayson Altieri, USA, retired; Mr. Daniel Armstrong; Dr. George H. Berghorn; Col Leonard Blascovich; Ms. Donna Budjenska; Dr. Andrew Burtch; Lt Col Paul Creed III; Dr. Susan Dawson; Mr. John Desmarais; Lt Col Brett Dolnick, USAF; Lt Col Wayne G. Fox; Maj Robert Frazier; Maj Robert Haase; Lt Col Jeremy Hodges, Col Jay Hone, USAFR, retired; Lt Col Seth Hudson; Mr. Milton F. Knight; Maj Erik Koglin; Cmdr (Dr.) David Kohnen, USN, retired; Ms. Nedra Looney; Lt Col (Dr.) Gerald Marketos; Capt Jerry Mason, USN, retired; Dr. Gregory Mathews; Maj (Dr.) Colleen McCormick; Lt Col John M. McDermott; Col Darin Ninness; Dr. Sarandis Papadopoulos; Maj Jill D. Paulson; Dr. Ryan Peeks; Lt Col Brenda Reed; Mr. John Salvador; Dr. Margaret Sankey; Mr. Thomas Shubert; Mr. Barry Spink; Mr. John Swain; Dr. Corbin Williamson; and Lt Col Charlotte Payne Wright. Finally, to my son, William, born during this project, I hope Daddy's book does you proud.

Abbreviations

AAF	Army Air Forces
AAFAC	Army Air Forces Antisubmarine Command
A&E	airframe and engine
ACER	Air Corps Enlisted Reserve
AFHRA	Air Force Historical Research Agency
AFNORTH	Air Forces Northern
AOPA	Aircraft Owners and Pilots Association
BLS	Barry L. Spink Collection
CAA	Civil Aeronautics Administration
CADS	Civil Air Defense Services
CAP	Civil Air Patrol
CAP-NAHC	Civil Air Patrol National Archives and Historical Collections
CAPNHQ	Civil Air Patrol National Headquarters
CAR	Civilian Air Reserve
CGAUX	Coast Guard Auxiliary
COMINCH	Commander in Chief, United States Fleet
CPTP	Civilian Pilot Training Program
ELJ	Earle L. Johnson Papers
EMP	electromagnetic pulse
FAC	forward air controllers
FEMA	Federal Emergency Management Agency
HHA	Henry Harley Arnold Papers
KKH	Kendall King Hoyt Papers
KTB	*Kriegstagebüch* (war diary)
LOC	Library of Congress

ABBREVIATIONS

NAA	National Aeronautic Association
NARA	National Archives and Records Administration
NDAC	National Defense Advisory Commission
OCD	Office of Civilian Defense
OEM	Office of Emergency Management
RPA	remotely piloted aircraft
sUAS	small unmanned aerial system
TIOH	The Institute of Heraldry
TWX	teletypewriter exchange services
WRHS	Western Reserve Historical Society
WTS	War Training Service

Chapter 1

Introduction

Each aviation task performed by civilians and their equipment helps release military planes and airmen for combat duty.
 —Office of Civilian Defense, *CAP Bulletin*, 8 May 1942

On 28 January 1943, Rep. Hatton W. Sumners (D-TX) spoke candidly before the House of Representatives about a group of volunteer civilian aviators whose services to the nation were little known. His "interest was aroused in this organization because of its demonstrated unselfish, self-reliant, willingness-to-do-something-about-it, fit-to-live-and-govern-in-a-free-democracy sort of spirit."[1] Responding to Sumners's comments, Rep. John M. Vorys (R-OH) noted how he himself flew active duty antisubmarine patrols with this civilian organization during the summer 1942 recess. Vorys praised their collective spirit "of self-reliance and resourcefulness, although in performing their flying missions their discipline and obedience meet military standards."[2] These were the men and women of the Civil Air Patrol (CAP).

Between the Office of Civilian Defense (OCD) and the War Department, the movement to militarize the civilian aviation community of the CAP into a fourth arm of the nation's defense unfolded over the first half of 1942. For 18 months from 1942 to 1943, these CAP civilian volunteers flying armed, light, privately owned aircraft operated an antisubmarine coastal patrol as part of the American military effort in the Battle of the Atlantic. This effort represented a prewar possibility turned wartime exigency, a product of multiple individuals bound by the idea that light aircraft and private citizens could serve a national defense purpose. In the context of this grand experiment, CAP members became the first American civilians to actively engage enemy forces in defense of the United States, proving the worth of auxiliary air services.

CAP's coastal patrol initiative developed from a critical need to stem the tide of German submarine operations inflicting heavy losses on coastal shipping. Begun as a subexperiment by the Army Air Forces in March 1942, CAP's effort commenced with meager resources and no practical experience in antisubmarine warfare. Senior Navy leaders cast a wary gaze upon the civilian undertaking and considered the entire endeavor to be problematic. Within weeks of operation, however, CAP's small effort demonstrated the discipline and military bearing desired by the Armed Forces. The Army embraced the CAP coastal patrol and eventually won over the Navy to boost the coastal antisubmarine deterrent and aerial convoy escort.

Through the initial months of the coastal patrol subexperiment, CAP demonstrated it could be semimilitarized to serve the needs of the Army Air Forces. The War Department thereafter provided the organization with additional funding and authorized wear of the Army uniform with distinctive insignia. By September 1942, CAP operated 21 bases flying approximately 423 privately owned, armed aircraft on continuous daytime coastal patrols from Maine to the Texas-Mexico border. The Army armed the light aircraft with bombs and depth charges while the aircrews received antisubmarine warfare training and operated under the Army Air Force's Antisubmarine Command. The success of the coastal patrol brought success for the grand experiment of CAP itself: private aviation could be organized and trained to serve the needs of the Army. In late April 1943, CAP transferred from OCD to the War Department, a move that subsequently ensured CAP's postwar survival.

Today, CAP operates as the civilian auxiliary of the US Air Force. The lone surviving element of the World War II–era OCD, the contemporary CAP is the civilian auxiliary of the US Air Force when performing Air Force–assigned missions and a congressionally chartered nonprofit corporation. CAP operations are oriented around three congressionally mandated tasks: emergency services, cadet programs, and aerospace education.[3] As a public service organization, CAP comprises uniformed male and female volunteers tasked with emergency response and disaster relief missions; providing diverse aviation and ground services; and working with youth development through the promotion of air-, space-, and cyberpower.

The corporation maintains and operates a fleet of 560 single-engine aircraft and an extensive very high-frequency and high-frequency communications network dispersed across all 50 states, Puerto Rico, the US Virgin Islands, and the District of Columbia. This force, larger than most of the world's air forces, equips the US Air Force with an operational and civil-military relations resource unlike any of the other uniformed services with the exception of the Coast Guard.[4] The membership of over 58,000 men and women includes approximately 10,000 aircrew members and 36,000 emergency responders trained to the standards of the Federal Emergency Management Agency (FEMA) and under the operational control of the First Air Force.[5] The volunteer service hours of CAP's members provide at least a four-to-one return on investment, thereby enabling the Air Force to increase training and liberate assets for deployment and greater operational use elsewhere.[6]

Following the discontinuance of CAP coastal patrol operations in late August 1943, however, this wartime story faded from national memory. Official CAP records have been lost or destroyed over the succeeding decades, hampering in-depth research and analysis of the organization's history. The CAP coastal patrol effort is unique in the organization's history as civilians received military weaponry and authority to attack enemy forces. This aspect of CAP's history frames discussion about the future employment of CAP in declared national emergencies or wartime contingencies.[7]

Scholarship about CAP remains limited. Several works appeared during and after the war discussing the coastal patrol effort and/or CAP's wartime effort. Notable among these are William Mellor's 1944 book, *Sank Same*; Andrew Ten Eyck's 1946 work, *Jeeps in the Sky: The Story of Light Planes in War and Peace*; and Robert E. Neprud's 1948 wartime CAP history, *Flying Minute Men: The Story of the Civil Air Patrol*.[8] Mellor's book resulted as an offshoot from a tasking by the Navy to write a series of articles publicizing the early antisubmarine efforts of CAP, the Coast Guard, and Coast Guard Auxiliary.[9] Likewise, Neprud's book represented a publicity-driven effort by the Army Air Forces to provide special recognition for CAP's wartime contributions.[10] Neprud's account of the CAP coastal patrol mirrored that in *Sank Same*, close enough in fact to warrant a successful lawsuit by Mellor.[11] In the official, somewhat more objective histories of the Army and Navy, CAP's coastal patrol operation is allotted a section in Samuel Eliot Morison's first volume in his voluminous *History of the United States Naval Operations in World War II*, and the operation is lightly referenced by editors Wesley Frank Craven and James Lea Cate in the Army Air Forces equivalent multivolume work, *Army Air Forces in World War II*.[12]

Unsurprisingly, over the ensuing decades fewer published histories explored CAP's wartime roles in any detail. Several books written by CAP members supplemented Mellor's research and Neprud's book within the pantheon of institutional history, in particular Charles B. Colby's *This Is Your Civil Air Patrol* and Frank Burnham's *Hero Next*

Door.[13] From the 1990s to the present, information about CAP's coastal patrol effort is frequently drawn from the previously published works.[14] A notable exception is Michael Gannon's 1991 best-seller, *Operation Drumbeat*. While Gannon's conclusions have been convincingly challenged by Clay Blair in his book *Hitler's U-Boat War: The Hunters*, Gannon did manage to introduce new material about CAP through Navy records at the National Archives.[15] In terms of original research, Louis Keefer's 1997 book, *From Maine to Mexico: With America's Private Pilots in the Fight Against U-Boats*, is an invaluable compilation of 275 oral histories from CAP coastal patrol veterans describing daily operational life and providing rich anecdotal accounts of the participants.[16]

The overarching problem hampering CAP history is the lack of accessible archival records. Of CAP's 21 bases, operational records exist only for one: Base No. 16, Manteo, North Carolina. Of CAP's wartime staff officers, the personal papers of its second national commander, located in the Western Reserve Historical Society in Cleveland, Ohio, are the only such files available for research. The transfer of CAP from OCD to the War Department shifted the organizational records, and holdings in the National Archives relating to CAP are limited to only a few small boxes. From 1941 to the present day, CAP's national headquarters moved five times and an unknown quantity of files were lost or destroyed in the process. Beginning in 1979 a small group of CAP members began an effort to save and organize the surviving historical files at CAP National Headquarters, Maxwell AFB, Montgomery, Alabama. A small percentage of these records were sent to the Air Force Historical Research Agency, also at Maxwell AFB, where they were microfilmed. For decades, these records constituted the bulk of available primary material on CAP.

As a result of the limited historical research about CAP and the coastal patrol operation, over the following 70 years, myth and lore transformed into hardened facts within the member ranks. New members, adult and cadet, learned from Neprud's history and/or abridged histories published by CAP National Headquarters.[17] All too frequently, any new discovery or reexamination of the history would be met with resistance should the information challenge the accepted, institutional narrative. Inadvertently aided by the lack of archival resources, professional historians would find themselves limited to the writings of Mellor, Neprud, and other snippets in published works. For over 70 years, no serious research effort attempted to challenge the orthodoxy, either from the CAP historians or the academic community.

The research into this study began in 2010 as the author began the laborious task of locating primary documentation. Wherever possible, the author sought sources, with documents and photographs being acquired from veteran CAP members, state archives, national archives, and even online auctions. In 2014 while serving as CAP's national historian, the author launched an effort to obtain office space on a US Air Force base to develop into an archive and storage facility for the material culture of the organization. In 2018, work began to move unprocessed caches of records and artifacts stored in closets, basements, and rental units in Washington, DC; Virginia; Maryland; Delaware; New York; Georgia; and Alabama to a 3,500-square-foot office space at Joint Base Anacostia-Bolling in Washington, DC. This space became the first home for CAP's heritage in the 79 years of the corporation. Named in honor of the first volunteer national historian, the Col Louisa S. Morse Center for Civil Air Patrol History made this study possible.

While the scope of this research focuses exclusively on the 18-month coastal patrol operation, readers should be aware of some of CAP's other military-funded operations, which echo the sentiments of the epigraph at the top of this chapter. CAP personnel participated in a variety of missions across the United States throughout the Second World War, with the most prominent among these including the following:

- Southern liaison patrol flights along the Mexican border from Brownsville, Texas, to the Arizona state line for the Southern Defense Command (3 October 1942–10 April 1944)
- Aircraft warning missions flown for the III Fighter Command to test the responsiveness of the service in Western Florida, Alabama, Mississippi, Louisiana, and Texas (1 August 1942–30 June 1943)
- Military courier service for the First, Second, and Fourth Air Forces carrying Army mail, aircraft parts, war materials, supplies, and personnel (27 August 1942–10 April 1944)[18]
- Tow target training operations flown for the First and Fourth Air Forces in support of antiaircraft gunnery training for the

Eastern and Western Defense Commands (1 December 1942–1 March 1945)[19]

- Missing-Aircraft Search Service flights at the request of the Army Air Forces to help locate missing military aircraft (10 May 1942–26 October 1945)[20]

While all these wartime missions warrant individual study, the genesis for them is found in the success of the coastal patrol mission. Hopefully, this history together with the resources of the Morse Center will serve as inspiration for future scholarship.[21]

CAP members participated in military endeavors as volunteers, motivated by patriotism, flying opportunities, and a commitment to service rather than as paid employment. The coastal patrol story is at times misunderstood or misinterpreted, treated with reverence but hampered by limited contextual insight and analysis as to its critical importance in the organization's maturation and survival. The coastal patrol participants are titled "flying minutemen" or "subchasers"— both superficial and misleading monikers that obscure the organizational development and policies of the operation, much of which deeply impacted CAP's wartime and postwar evolution.

This book attempts to achieve four main objectives: First, inform and educate readers on the overall particulars of CAP's coastal patrol effort. Second, examine how CAP's relationship with the Army and Navy's domestic antisubmarine operations resulted in the semimilitarization of wartime CAP. Third, using the overall study of the coastal patrol operation, draw a series of principles to form a doctrinal basis for the discussion of future uses of civilian Airmen for domestic military purposes in time of war. A fourth objective is found in the endnotes for each chapter. These are intentionally extensive to provide future scholars with the sectional charts required to navigate the early history of CAP and pioneer new routes for research and exploration.

INTRODUCTION | 7

Notes

(Endnotes will be a mix of shortened form and discursive entries. For those endnotes presented in short form, the full resource is listed in the bibliography.)

Epigraph. "Need for Aviation Resources," Office of Civilian Defense (OCD), *CAP Bulletin* 1, no. 15 (8 May 1942), Folder 6, Box 6, Earle L. Johnson Papers (ELJ), Western Reserve Historical Society, Cleveland, OH (WRHS)

1. Speech of the Honorable Hatton W. Sumners of Texas in the House of Representatives, 28 January 1943, "The Civil Air Patrol—Patriotic Service of Great Value" (Washington, DC: Government Printing Office [GPO], 1943), 3.

2. John M. Vorys, quoted in above, Sumners, "The Civil Air Patrol—Patriotic Service of Great Value," 7. Vorys flew as an observer (and possibly as a pilot) out of Coastal Patrol Base No. 14, Panama City, Florida, in August 1942, making him the only member of Congress to actually fly coastal patrol missions with CAP during the war. Keefer, *From Maine to Mexico*, 330–31; John M. Vorys to Bennett C. Clark, 25 September 1942; and John M. Vorys to Earle L. Johnson, 25 September 1942, Barry L. Spink Collection (BLS), CAP National Archives and Historical Collections, Col Louisa S. Morse Center for Civil Air Patrol History (CAP-NAHC), Joint Base Anacostia-Bolling, Washington, DC.

3. The incorporation legislation for CAP listed the objects and purposes of the corporation as "to provide an organization to encourage and aid American citizens in the contribution of their efforts, services, and resources in the development of aviation and in the maintenance of air supremacy, and to encourage and develop by example the voluntary contribution of private citizens to the public welfare" and "to provide aviation education and training especially to its senior and cadet members; to encourage and foster civil aviation in local communities and to provide an organization of private citizens with adequate facilities to assist in meeting local and national emergencies." Beginning in the 1950s, these purposes began to be expressed in simplified form as three program areas which remain in use today: cadet programs, aerospace education, and emergency services. Act to Incorporate the Civil Air Patrol, Public Law 79-476, US Statutes at Large 60 (1946): 346–47, codified at US Code, Title 36, chap. 403 (2000).

4. Congress established the US Coast Guard Auxiliary on 23 June 1939 and established the Coast Guard Reserve, designated as the US Coast Guard Auxiliary (CGAUX), on 19 February 1941. The CGAUX's three missions are to promote and improve recreational boating safety; provide trained crews and facilities to augment the Coast Guard and enhance safety and security of the nation's ports, waterways, and coastal regions; and support Coast Guard operational, administrative, and logistical requirements. US Department of Homeland Security, United States Coast Guard Auxiliary, "About the Auxiliary," 31 August 2019, http://cgaux.org/about.php; and Tilley, *United States Coast Guard Auxiliary*.

5. Civil Air Patrol National Headquarters (CAPNHQ), "2019 Fact Sheet," https://presspage-production-content.s3.amazonaws.com/uploads/1913/capfactsheet201911-581843.pdf?10000; and John W. Desmarais, Headquarters, CAP, Director of Operations, Maxwell AFB, AL, to the author, e-mail, 25 August 2019.

6. CAP, *Civil Air Patrol–US Air Force Auxiliary 2016 Financial Report*, 27, 35.

7. This discussion is oriented exclusively on the emergency service aspects of the corporation. CAP's aerospace education and cadet programs are both elements of immense value in the contemporary and future development of airpower in the

United States but outside the scope of this study. The cadet program itself warrants a detailed study extending from World War II to the present day in regard to the role and impact of CAP on the recruitment and retention of personnel within not only the Air Force but also the entire Armed Forces.

8. Mellor, *Sank Same*; Ten Eyck, *Jeeps in the Sky*; and Neprud, *Flying Minute Men*.

9. J. L. Tinney to District Coast Guard Officer, Fourth Naval District, memorandum, subject: Mellor, William B., Jr., Philadelphia Record—Assignment of, 13 March 1944; William B. Mellor to Kendal K. Hoyt, 23 March 1944; and Earle L. Johnson to William B. Mellor Jr., 24 March 1944, Reel 38920, Air Force Historical Research Agency, Maxwell AFB, AL (AFHRA).

10. Neprud, interview.

11. Mellor learned of the similarities between *Sank Same* and *Flying Minute Men* in February 1950. He agreed to settle out of court, and CAP paid $525 in damages. William B. Mellor Jr. to Duell, Sloan and Pearce Inc., 15 February 1950; Franklin E. Welch to Charles A. Pearce, 21 February 1950; Franklin E. Welch to Judge Advocate General, memorandum, subject: "Flying Minute Men," 29 March 1950; Robert E. Neprud to Freeman C. Bishop, 3 April 1950; John M. Webster to Franklin E. Welch, 4 April 1950; Franklin E. Welch to Wallace D. Newcomb, 5 April 1950; Wallace D. Newcomb to Franklin E. Welch, 10 April 1950; Franklin E. Welch to Charles A. Pearce, 11 April 1950; Andre Maximov to Franklin E. Welch, 19 April 1950; and George W. Witney to Franklin E. Welch, with attachment, "Points of 'Similarity' between 'Flying Minutemen' and 'Sank Same,'" 21 July 1950, BLS, CAPNHQ.

12. Morison, *Battle of the Atlantic*; and Craven and Cate, *Plans and Early Operations*.

13. Colby, *This Is Your Civil Air Patrol*; and Burnham, *Hero Next Door*.

14. Two such examples are Hickam, *Torpedo Junction*, and Offley, *The Burning Shore*.

15. Gannon, *Operation Drumbeat*; and Blair, *Hitler's U-Boat War*.

16. Keefer, *From Maine to Mexico*.

17. CAP, *Civil Air Patrol Manual: An Introduction*, 11–20; *Civil Air Patrol Manual 50-1, Introduction to Civil Air Patrol*, January 1958, 1975, and 1978 revisions; and Civil Air Patrol Pamphlet 50-5, *Introduction to Civil Air Patrol*, April 2013.

18. Harry H. Blee to Earle L. Johnson, memorandum, subject: Weekly Operations Report, 30 June 1944, Reel A2292, AFHRA; H. A. Craig to Earle L. Johnson, memorandum, subject: Discontinuance of Certain Civil Air Patrol Activities Carried Out at Government Expense, 4 April 1944; and Earle L. Johnson to Assistant Chief of Air Staff, Operations, Commitments, and Requirements, memorandum, subject: 1st Ind., Discontinuance of Certain Civil Air Patrol Activities Carried Out at Government Expense, 6 April 1944, Binder "Legal Status, Administrative Concepts, and Relationship of the Civil Air Patrol, 1941 to 1949," CAP-NAHC.

19. L. O. Ryan to Earle L. Johnson, memorandum, subject: Discontinuation of towing and tracking missions performed by the Civil Air Patrol, 11 January 1945, Reel 38918, AFHRA; "CAP Relieved of Perilous Job," *New York Daily News*, 22 January 1945, 6; and "AAF to Tow Own Targets," *Cincinnati Enquirer*, 22 January 1945, 7.

20. On 10 May 1942, the Army Air Forces first turned to CAP to assist in locating a missing aircraft. The following year in May, Army Air Forces Regulation No. 20-18 authorized CAP to "engage in such other activities in behalf of the armed forces or of federal, state, or municipal authorizes, or persons or corporations engaged in the war effort as are deemed advisable by the National Commander and are within policies established by the Commanding General, Army Air Forces." Subsequently, CAP National Headquarters formally organized a Missing-Aircraft Search Service funded by the military to assist in locating lost military aircraft as requested by the Commanding General of the Army Air Forces. The dates listed are based on a statistical

summary from 30 November 1945. War Department, Headquarters Army Air Forces, Regulation No. 20-18, Organization—Civil Air Patrol, 25 May 1943; document, "Civil Air Patrol," 8 June 1945, Binder "Legal Status, Administrative Concepts, and Relationship of the Civil Air Patrol, 1941 to 1949"; CAPNHQ, Summary of CAP Missing-Aircraft Search Service, 30 November 1945, BLS, CAP-NAHC; CAPNHQ, Operations Memorandum No. 25, Procedure for Missing-Airplane Search Missions, 15 April 1944; CAPNHQ, Operations Directive No. 44, CAP Missing-Aircraft Search Service, 15 April 1944 and 1 July 1944; CAPNHQ, Summary of CAP Missing-Aircraft Search Service, 30 December 1944; and Barney M. Giles to Commanding Generals, All Air Forces, All Independent and Subordinate AAF Commands, Commanding Offices, all AAF Bases (in Continental United States), memorandum, subject: Utilization of Civil Air Patrol in Searches for Lost Aircraft, 15 June 1944, Reel A2992, AFHRA.

21. Detailed information about the tow target, courier, and missing-aircraft search and assorted other missions are almost exclusively limited to archival sources. The three primary repositories of such materials are the CAP National Archives and Historical Collections at the Col Louisa S. Morse Center for CAP History, the Air Force Historical Research Agency, and the Earle L. Johnson papers at the Western Reserve Historical Society. The best overall secondary source discussing all the CAP wartime missions is Neprud, *Flying Minute Men*. For specific discussion of the Liaison Patrol, see Ragsdale, *Wings Over the Mexican Border*.

Chapter 2

Origins of the Civil Air Patrol

Following the studies and conversations which have taken place on the question of mobilization of civilian aviation potentiality for the joint benefit of national defense and civilian aviation, it has been decided that this Office shall immediately undertake the formation of a volunteer national organization of pilots, mechanics and other aviation personnel to be known as the "Civil Air Patrol."

—Fiorello LaGuardia, 29 September 1941

CAP's overall origins, like its coastal patrol effort, trace back to Germany. In autumn 1936, New Jersey state aviation director, Gill Robb Wilson, boarded the German airship *Hindenburg* at Naval Air Station, Lakehurst, for passage to Frankfurt. Born in Clarion County, Pennsylvania, the son of a Presbyterian minister, Wilson grew up reading about and witnessing a flurry of new technological wonders, including heavier-than-air flight. Wilson did not actually see his first aircraft in flight until 1911, the year he entered Washington and Jefferson College. After graduating in 1915 he entered the Western Theological Seminary in Pittsburgh but left in the fall with his younger brother, Joseph Volney, to drive ambulances for the French army. On the last day of his voyage to France, a German U-boat unsuccessfully attacked Wilson's ship.[1]

As a member of the American Ambulance Service, Wilson witnessed the bloodiness and suffering in and around the Vosges Mountains and in the Verdun sector. Desiring to be an airman, he spent his free time at French airfields studying operations and making friends. After finishing his commitment to drive ambulances, Wilson transferred into the French Air Force (*Armée de l'Air*) and joined his brother as a bomber pilot, with Joseph then a member of the Lafayette Flying Corps. Wilson flew with *Escadrilles* Br. 66 and Br. 117 until the death of his younger brother in October 1918, at which point he transferred to Joseph's unit, the US Air Service's 163rd Bombardment Squadron.[2] Until the armistice, Wilson served as the unit's operations officer and worked to prepare the unit for combat.[3]

Discharged as a first lieutenant in 1919, Wilson entered the reserves and returned to Western Theological Seminary to complete his studies. After graduating in 1920, he moved to New Jersey to pastor the Trenton 4th Church. He maintained his interest and involvement in aviation, both military and civilian. As a member of the American Legion, Wilson "hammered at the necessity of aviation as military power. Out in the sticks I pounded for airports on the thesis that a community without a field would become like a town without a railroad station."[4] Elected national chaplain of the American Legion in 1927, he traveled the country and expanded his network of acquaintances in faith and aviation.[5] Three years later he became the first state director of aviation for the New Jersey Department of Aviation and was active in the 119th Observation Squadron of the state National Guard. In these roles, Wilson experienced firsthand the utility of light aviation for civilian and military use.[6]

On his 1936 trip to Germany Wilson met with various acquaintances and gathered firsthand knowledge of the Third Reich's air development. Upon his arrival Wilson received a letter of welcome from the German Ministry of Aviation (*Reichsluftfahrtministerium*). The following day Wilson met with a major from the Wehrmacht, assigned to arrange interviews, visits of aircraft plants, aviation demonstrations, or anything else the American official desired. Numerous aviation officials freely shared every aspect of Germany's aviation progress to date. Wilson was particularly intrigued by the Air Ministry's interest in "a mere soaring competition among youth clubs." Observing German youth launching and flying gliders and engaged in aspects of aviation education, it soon dawned on him that "here was a military indoctrination of airmen on the broadest possible basis. Its fruition must be some five years away, but it would be a massive harvest when it came."[7] One evening over wine, Wilson's guide mentioned his service in the First World War as a submarine officer, proclaiming "your East Coast is the best hunting ground in the world for submarines."[8] Wilson recognized the stirring winds of war and needed time to think.

Back in the United States, Wilson contemplated what American aviation could do to catch up to Germany's. His thoughts centered on what role light aviation, both aircraft and private pilots, could play in national defense. Wilson consulted with various leaders in civil aviation. Piloting his own light aircraft, he flew out along the New Jersey coast to see if low-altitude observation flight could identify objects.

Convinced light aviation would play a valuable role in the inevitable future conflict, Wilson wrote to Rear Adm Cary T. Grayson, chairman of the American Red Cross, and proposed a Red Cross auxiliary aviation corps. With trained, patriotic pilots engaging in assorted emergency missions such as transporting medical supplies, evacuating injured citizens, or patrolling flood-prone areas, a Red Cross auxiliary aviation corps would build up valuable experience using light aircraft, and this experience in turn would be invaluable for cooperation with the armed forces in time of military crisis. He received no reply.[9] A possible explanation may be that National Guard aviation units were performing these roles at the time of Wilson's proposal, something he would be familiar with as a guardsman himself.[10]

Figure 1. Gill Robb Wilson, intellectual founder of the Civil Air Patrol.
(Photograph courtesy of the Col Louisa S. Morse Center for CAP History, hereafter Morse Center.)

Other private citizens independently conceived the idea of employing civilian aviation for national defense. In Toledo, Ohio, a group of 11 pilots incorporated the nonprofit Civilian Air Reserve (CAR). Milton Knight, secretary of the Libbey-Owens-Ford Glass Company, had conceived the idea. A graduate of Yale Law School, Knight was an avid sportsman pilot and yachtsman. His thinking behind CAR was to stimulate private flying for those pilots interested in doing more than just circling about airports. If trained and focused on perfecting their technique, these aviators could make themselves valuable as prospective military pilots and aid national defense. On 12 November 1938, Milton Knight; his brother and fellow pilot Edward F. Knight; Thomas B. Metcalf, owner of the Metcalf Flying Service at Toledo's Municipal Field; and Clarence A. Carson, safety director at Libbey-Owens-Ford, signed the articles of incorporation.[11]

The incorporation articles listed three primary objectives to guide CAR's efforts: develop interest in aviation; offer trained and student pilots the incentive and opportunity to keep up their training and experience; and mature a secondary or potential reserve of pilots, mechanics, and other personnel "who can be of service to their country in times of national emergency."[12] A fourth objective sought "to endeavor to interest the Federal and State governments in and to expand the organization of the Civilian Air Reserve in all States, and to seek and obtain governmental support and assistance in the further development of the purposes of this organization as part of a program of national defense."[13] For peacetime, the CAR envisioned its services of value in the event of fire, flood, or other situations where their aircraft, radios, and photographers could assist with the Red Cross or other relief organizations.

CAR's organizational model was self-described as "semi-military," and the organization provided a clear picture of what could become of civilian aviation for national defense purposes. Copying the Army Air Corps' structure, squadrons were organized with officers with military titles and ranks together with flights and enlisted ranks and aircrew. States would be organized with wings subdivided into groups, squadrons, and flights. Members 16 years of age and up could join once they passed membership examinations. CAR personnel wore uniforms of gray pants and gray Army-style shirts with insignia on the sleeves and shoulders and rank the "same as those adopted in the U.S. Army regulations."[14] CAR's volunteer pilots and aircraft practiced formation flying, navigation, meteorology, radio communication, aerial photography, theory of flight, and aircraft and engine maintenance to augment the nation's air defense forces should the government request their services.[15] From the original Toledo unit, additional CAR units formed from 1939 to 1941 in several other states, including Colorado, Connecticut, Florida, Maine, Massachusetts, New York, and Pennsylvania.[16] While relatively small, the CAR demonstrated what volunteer civilian airmen could accomplish for national defense purposes.[17]

Figure 2. Milton Knight, founder of the Civilian Air Reserve, seen here in 1944 as a lieutenant in the Naval Reserve aboard the destroyer escort USS *Holton* (DE-703). (Photograph courtesy of Milton F. Knight.)

Wilson's predictions of war proved all too accurate as war commenced in September 1939 in Europe. The American home front remained isolationist, and the organization of civilian resources remained a local or private matter. As with other key components of the nation's security apparatus, the invasion and fall of France served as the catalyst for sparking domestic mobilization. On 25 May 1940, Pres. Franklin D. Roosevelt established the Office of Emergency Management (OEM) by administrative order within the Executive Office of the President. Tasked with coordinating the nation's defense program, OEM served as "the incubator for many defense and war organizations."[18] Four days later following a declaration of an unlimited national emergency, Roosevelt reestablished the First World War–era Council of National Defense. He appointed a National Defense Advisory Commission (NDAC) to advise and coordinate the nation's industrial infrastructure for defense production. As NDAC became inundated with problems and inquiries from state governments, it established the Division of State and Local Cooperation on 31 July. Directed by

Frank Bane, executive director of the Council of State Governments, the division served as the channel of communications between the Council of National Defense, NDAC, and the various state defense councils. Bane's division also shared information with state and local councils about new developments and changes in the national defense program.[19]

In August, NDAC sent all state governors guidance on "State and Local Cooperation in National Defense." The commission suggested to the governors that, if desirable, "a state council of defense be created, and that such councils in turn, guide and assist in the formation of councils of defense in the local subdivisions of the State whenever the need becomes apparent." Suggested functions for the state councils of defense involved "advising the governor on problems arising with respect to the (1) integration of governmental programs for defense; (2) adjustments or arrangements necessary for prompt assimilation of such programs by the administrative establishment; [and] (3) proper coordination between the activities of government and private agencies coordinating in the defense effort."[20] Governors and their respective state defense councils would have considerable latitude to implement ideas and suggestions.

After the Second World War broke out in Europe, voices concerned about American home defense appeared in popular media. Such sentiments bore the faint echoes of the Preparedness Movement before American entry into the First World War.[21] Roosevelt's reestablishment of the Council of National Defense added to the chorus of voices of Americans wondering "what about us?" A year later, Americans listened to journalist Edward R. Murrow's nightly radio broadcasts reporting on the Luftwaffe's bombing campaign against London. Murrow's reports on the Blitz further stirred Americans to contemplate if, how, or when they too might be subject to nightly visits by enemy bombers. Civic leaders began to ask what the federal government intended to do to safeguard the home front from aerial bombardment as then visited on Europe.[22] Civil and federal officials in turn asked what resources existed among the population for defensive purposes. In July 1940, Knight began to schedule a national convention to establish a federally legislated, nationwide Civilian Air Reserve. That October, the Department of Commerce's Civil Aeronautics Administration (CAA) held a conference for its new Aeronautical Advisory Council. The 18-member body, which included, among others, Wilson and publisher Guy P. Gannett, president of Guy Gannett Publishing, appointed Knight to chair a subcommittee to plan a national program akin to his CAR organization.[23]

On 21 August 1940, Wilson, now president of the National Aeronautic Association (NAA), wrote to Audley H. F. Stephan, chairman of the recently established New Jersey Defense Council, about employing civilian aviation for national defense purposes. He explained to Stephan his previous outreach to the Red Cross and how the organization should establish an auxiliary aviation corps. The war's spread, Wilson concluded, made it "inevitable that the activities of civilian aviation will be curtailed in the interest of military aviation" as there was insufficient room for the simultaneous development of civil and military facilities.[24]

Wilson remained optimistic, believing national defense and civil air growth could coexist. As he explained to Stephan, "Now there are numbers of thoroughly capable pilots who are not and could not be utilized in the military service." The pilots, he continued, "are eager to do anything in their power and to utilize their experience for the national defense but they do not know how to proceed and no constituted agency of government has given them any light on the subject, nor offered them any opportunity."[25] Under the aegis of the state defense councils, a corps of these civilian aviators could be vetted, trained, and organized for observation and ferrying work. When military operations came to a given area, these individuals could assist in carrying messages or passengers and provide useful information. Wilson did not want to launch such an organization in New Jersey "unless it has a specific utility" and the backing of the military and the Red Cross. Wilson proposed that if the Council of National Defense was interested, he would use his presidency of the NAA to call up key individuals to launch a civil air organization in the space of three months. If Stephan wanted such an organization in New Jersey as a trial balloon, Wilson would gladly organize the unit.[26] Stephan, in turn, forwarded Wilson's letter to Bane for NDAC advice, who agreed to discuss the matter with other appropriate federal agencies.[27]

In September 1940, President Roosevelt ordered the National Guard—whose aviation units furnished the Army Air Corps with 29 observation squadrons—to active duty.[28] This action drastically reduced the aviation resources available to state governments. That same month, the Aircraft Owners and Pilots Association (AOPA), taking cues from CAR, launched its own uniformed organization, the Air Guard. Formed in consultation with and receiving assistance from the Army Air Corps and the Adjutant General's office, the AOPA initiative began as an extension of an emergency pilot corps of private flyers available to assist local communities nationwide with domestic, natural disaster–variety emergencies. An extension of this formed the basis for an "auxiliary to the Army reserve pilots." As designed, the Air Guard consisted of three divisions, known as the One, Two, and Three Star courses. The former consisted of a classroom period, mirroring the Army extension courses for second lieutenants. The Two Star course focused strictly on controlled flying and developing flying skills. The third portion involved those pilots with instrument and night flying experience who would be considered advanced rated. Those unable to advance beyond the Two Star course could qualify for a specialist's rating in aerial photography, observation, or other skill sets. AOPA planned for the Air Guards to have uniforms and special insignia for both aircraft and personnel. While directly managed by AOPA, the organization's goals included official US Army recognition together with direct observation and supervision of the effort to become a civilian auxiliary of the Army Air Corps.[29]

In a February 1941 report to President Roosevelt, New York City mayor Fiorello H. LaGuardia—a former First World War bomber pilot—recommended creating a home defense organization among the population and training ordinary citizens to meet the threat of air or naval attack on American cities.[30] His message spurred civil aviation advocates to action. On 22 April, Thomas H. Beck, president of the Crowell-Collier Publishing Company and a committed promoter of civil aviation, hand delivered a letter to Roosevelt that outlined a "plan for increasing defense and war-consciousness and for enlisting the youth of the United States in aviation." Among the elements of his plan, Beck recommended the Bureau of Education provide textbooks and model airplanes to elementary schools, provide a glider for every high school to teach glider flying—likely a nod to the fruitful German youth effort—and organize all Civilian Pilot Training Program (CPTP) students not accepted into the armed forces. The latter would form a youth aviation patrol on the nation's borders, flying aircraft for patrol, observation, and radio communication training.[31]

The first weeks of May 1941 saw a flurry of legislation introduced to leverage civil aviation for national defense purposes. No fewer than four bills in the House and one in the Senate all sought to establish a CAR.[32] The first of these, H.R. 4664, introduced by Congressman

John E. Rankin (D-MS) on 6 May 1941, would establish a CAR of qualified civilian pilots, mechanics, ground-crew members, and aircraft owners, organized and trained by the War and Navy Departments, in the interest of national air defense. The CAR personnel would be subsidized for services in war exercises and training in accordance with War and Navy Department regulations and fees. The bill also included provision for Army and Navy officers to supervise the provision and installation of "necessary war equipment" on private aircraft, including bombing racks, bombs, machine guns, and light armor.[33]

In providing its position on the legislation, the CAA supported a CAR for periods of emergency to carry out nonmilitary flying assignments. The CAA, however, expressed a belief that CAR service should be voluntary and the organization financially self-supporting. Most of all, the CAA doubted "the advisability of training members of the civilian air reserve in military operations of combatant character," preferring the civilian aviators engaged in noncombatant roles such as ferrying aircraft, courier service, or disaster relief work. Donald H. Connolly, administrator of civil aeronautics, informed Commerce Secretary Jesse H. Jones that since the proposed legislation had certain defense significance, the CAA would not object to the bill if the War and Navy Departments approved of the matter.[34]

On 20 May, Roosevelt issued Executive Order 8757, establishing the Office of Civilian Defense (OCD). Roosevelt tapped LaGuardia to serve as director and tasked OCD with coordinating federal civilian defense activities with those of state and local governments.[35] With civil aviation's role unclear, Beck shared his Roosevelt letter with Gannett. In the first week of June, Gannett shared Beck's plan with LaGuardia.[36] Recognizing potential, on 12 June, LaGuardia appointed Beck, Gannett, and Wilson as a committee.[37] LaGuardia tasked the men to "formulate plans and submit suggestions to me as to the enrollment of private planes, owners and pilots and suggestions for their use in connection with the Civilian Defense program."[38] The nation's civil aviation resources around this time numbered more than 2,500 landing fields, several hundred flying schools, approximately 100,000 private pilots, and a comparable number of student and ex-pilots.[39]

Simultaneously in New Jersey, Wilson worked to finalize efforts for the launch of the New Jersey Wing of the Civil Air Defense Services (CADS). His plan for a "Civil Air Guard" had reached Gov. Charles

Edison in late March. Wilson described the membership as purely voluntary, without expense to the state and without "any complicated machinery to make it other than it is—a simply [sic] and useful instrument to employ the capacity of willing and able Americans to serve the state and nation within the limitation of their civil status."[40] Days later Wilson wrote the governor again, this time requesting approval or disapproval of the plan to the state defense council. "Speed is essential," noted Wilson, "because the New Jersey pattern is to be used as the national pattern."[41] Within two weeks, Edison recommended its implementation by the council.[42]

Throughout April and May, Wilson assembled a Civil Air Defense Committee to finalize the planned launch and recruitment of members, and the State Defense Council approved the final plan on 13 June. Wilson's introductory proposal called for recruiting 220 pilots and an equal number of mechanics with hundreds of additional personnel including photographers, radio operators, and office staff. After training in technical flying, first aid, and familiarization with the state's topography, the organization would be available for use by civilian defense.[43] "One purpose is to develop an esprit de corps in civil aviation," proclaimed Wilson, "to meet the problems and preserve the integrity of civil aviation through organized cooperation in civil defenses."[44]

Actual recruiting for New Jersey's CADS began on 10 July. As designed, the program organized the state's civil aviation resources for effective cooperation with the military and civilian defense forces. Other objectives included developing morale and discipline within civil aviation, organizing the accumulated knowledge of civilian pilots, providing aerial transportation, and aiding defense purposes through observation and guarding of public works and industrial areas. The CADS used the organizational structure of the Army Air Corps, with members holding the rank of cadet until qualified for officer ranks.

The New Jersey Wing of CADS consisted of three groups, each with two squadrons composed of three or more flights and assigned to different sections of the state. Training involved 200 hours of classroom instruction on topics ranging from air regulations, navigation, and meteorology to power plants, parachutes, and first aid. The new organization featured specialized insignia for wear by male and female members but otherwise had no designated uniform. By fall, the

New Jersey Wing numbered over 600 members and 157 privately owned aircraft.[45]

On 17 June, Beck, Gannett, and Wilson met in New York City to draft their plan. Kendall K. Hoyt, NAA's manager, also attended and took minutes.[46] On 25 June, they presented their proposed plan for Civil Air Defense Services to LaGuardia. The proposal had two objectives:

1. The immediate organization of available civil air resources
2. The ultimate civil development essential to any sound foundation for airpower[47]

The former sought to organize civilian aviators ineligible for military service but potentially eligible for auxiliary service in a national emergency. Properly trained and organized along the lines of the Army Air Corps, a disciplined civil air component could "patrol against flight over restricted areas of industry, potable water supply, cities, arsenals or other areas where sabotage or the gathering of information from the air might interfere with national defense." More broadly, this civil air component could "provide an organized service available for use in emergency disaster when military aircraft and personnel were otherwise engaged." All personnel would serve on a voluntary basis, with federal funding provided only for fuel, lubrication, and maintenance; the states would absorb all remaining costs.[48]

For the latter objective, the three planners advocated for aviation education. The proposal recommended providing funds for schoolteachers to attain private pilot licenses and establishing flight scholarships for high school students. Developing and fostering aviation-minded youth would be a key tenet for this proposed program. The proposal's lone visual graphic, "The Pyramid of Civil Air Power" (see appendix G), rested upon two tenets: indoctrination of youth through the public schools of the United States and savvy use of press and radio to impress on the public the virtues of being "air-minded."[49]

This concept of "air-mindedness" was not unique at the time. In his influential work, *The Command of the Air* (*Il dominio dell'aria*), Italian airpower theorist Gen Giulio Douhet linked the strength of a nation's military air force to its civilian aviation industry. "Civil aviation employs planes, trains pilots and maintains them in active service, and makes use of various aviation accessories—all means directly utilizable by the organs of national defense," wrote Douhet.[50] He wrote of how civil aviation needed to receive federal support and financial aid, including funding research and development. Civil aviation, in turn, had a responsibility to develop air-mindedness among the public. This would include developing a pool of pilots and mechanics and the promotion of the economic, social, and military benefits of airpower through air races, air shows, and exhibition flights.[51] Although it is unclear if Beck, Gannett, or Wilson had read, much less knew of, Douhet, they undoubtedly had absorbed the writings of the late Brig Gen William "Billy" Mitchell, which promulgated the same ideas linking a nation's military airpower and its civil aviation community.[52]

Concurrent with development of the CADS plan, the War Department weighed the fate of civil aviation's role on the home front. On 20 June, Secretary of War Henry L. Stimson wrote to Secretary of Commerce Jones expressing concerns about the regulation of nonscheduled civilian flying, which posed a security concern regarding the protection of all national defense activities. Stimson had received suggestions that the only effective means to eliminate any possible threat from civilian flying was to ground all aircraft except military, airlines, and recognized aviation agencies. The War Department, however, was reluctant to recommend airspace reservations or restrictions on civilian pilots and aircraft. Stimson explained departmental opinion that "national interest in aviation should be stimulated and not stifled." He instead suggested that the Commerce Department might want to investigate the loyalties and activities of civilian flyers and perhaps to organize those civilian flyers of "unquestionable loyalties" into an "'Air Vigilante' or Civilian Air Secret Service for the purpose of surveillance and counter-espionage."[53] Jones considered such an effort viable and reported the CAA would proceed immediately to create the organization.[54]

Perhaps without CAA knowledge, LaGuardia and OCD sent the CADS plan to the War Department. As an aviation matter, the plans reached the desk of Maj Gen Henry H. "Hap" Arnold, chief of the Army Air Forces, for review. One of the Army's first pilots, Arnold's career frequently crossed paths with private aviation. He directly facilitated numerous experiments using light aircraft in a variety of roles including missing aircraft searches, spotting forest fires, courier duties, or tactical airlift. A true believer in the cause for an independent air force, Arnold sought opportunities to educate the public and political leaders about Army aviation, either through his personal writings or in his official military capacity. To build up the Army Air Corps as war clouds gathered in Europe, Arnold fully grasped the critical role of scientific research in the civilian aviation industry in advancing military aircraft and weaponry. He also needed large numbers of trained, competent aviation personnel.[55]

During his tenure as chief of the Army's aviation community, Arnold had observed the CAA's effort to grow civil aviation through education and training. In December 1938, President Roosevelt had announced the creation of the Civilian Pilot Training Program. Brainchild of CAA chair Robert H. Hinckley, CPTP planned to train 20,000 civilian pilots annually. By using classrooms in US colleges and universities in partnership with nearby flying schools, male and female college undergraduates of all races—albeit in segregated flying schools—could receive 72 hours of ground school and from 35 to 50 hours of flight instruction. Born in the Great Depression, CPTP would also provide a much-needed boost to the private aviation industry for economic gain while creating a pool of potential military pilots for war preparedness.

Hinckley believed in fostering air-mindedness among the nation's youth. Like Wilson, he recognized the influence of the Reich Ministry of Aviation in the German education system. He considered CPTP a means to increase aviation's technological and intellectual influence on American society with an eye to more peaceful, prosperous times. By the summer of 1941, however, CPTP's civilian program increasingly shifted to a more military-oriented effort. No less than Arnold himself had concerns about the quality of the CAA-administered aviation training and the negative impinging of CPTP on resources needed for the Army Air Corps' training efforts.[56]

Within this context, Arnold reviewed the proposed CADS plan and shared his thoughts with LaGuardia. On 19 July 1941, Arnold told him about the numerous requests the Army had received from civilian pilots interested in ferrying aircraft. Many of these pilots, however, were not suitable for the work, due to age, education, or marital status, among other reasons. Arnold suggested LaGuardia and OCD establish an agency to catalog and classify all male pilots in the country unfit for military service or for ferrying aircraft "so that, if conditions later warranted, we could call upon you for quantities of personnel with definite abilities on certain types of aircraft who could be used for the ferrying of military aircraft."[57]

After reviewing the CADS plan, Arnold recognized a potential beyond the mere ferrying of aircraft. In a 28 July letter to Brig Gen Lorenzo D. Gasser, the War Department representative to OCD's Board of Civilian Protection, Arnold shared his thoughts about the

CADS plan. He referenced the earlier CAA-sponsored CAR plan and a previous unwillingness of the Air Corps to express an opinion on the matter. The air chief affirmed that "the organization of the *existing* private flying resources is highly desirable from a National Defense standpoint" (emphasis in original). He believed OCD's new planned organization could be "most advantageously put into effect" but that it "must be definitely prepared to stand on its own feet" without military assistance, opposing any effort to increase private pilot activity or aircraft manufacturing at the expense of the purely military effort. Should LaGuardia's plan come to fruition, Arnold deemed it essential that a "non-professional character be preserved, and that the personnel involved be strictly limited to military non-effectives."[58]

LaGuardia appointed his aviation aide, Maj Reed G. Landis, to turn the CADS proposal into an actionable plan.[59] Appointed in late July, Landis and LaGuardia shared an aviation background forged in World War I. The only son of Major League Baseball Commissioner Judge Kenesaw Mountain Landis, the Ottawa, Illinois, native began his military career in the National Guard with the 1st Illinois Cavalry before joining the Aviation Section of the Signal Corps in 1917. As a fighter pilot in France, he shot down nine enemy aircraft and one balloon, receiving the British Distinguished Flying Cross and the American Distinguished Service Cross. In late September 1918 he took command of the 25th Pursuit Squadron until the armistice.[60] After the war, Landis was instrumental in organizing the National Association of State Aviation Officials, serving as its president and as the first chairman of the Illinois Aeronautics Commission. At the time he was "lent" to OCD, Landis was in Chicago, serving as regional vice president of American Airlines.[61] With a solid background in both military and civil aviation, Landis brought OCD a vast professional network and wealth of aviation knowledge.

In mid-August 1941, an informal board of three Air Corps officers studied the various CAR, CADS, and Air Guard plans and the proposed methods of regulating and controlling the flight of civil aircraft to reduce possible sabotage and subversive activities.[62] The board consulted with several civilian experts, notably Wilson, Landis, Hoyt, and John B. Hartranft, executive secretary of AOPA. The CAR's founder, Knight, was not consulted. This apparent omission may have been due to Knight being busy as a company vice president answering inquiries about specially fabricated glass for defense use, leaving no time to participate in CAR plans or future civil aviation issues.[63] Curiously, the officers listed the AOPA's Air Guard as the only national "quasi-military reserve" of civilian pilots or other airmen then in operation. The Florida and New Jersey flying organizations sponsored by the respective state defense councils, received only passing reference.

The informal board concluded that, properly organized and trained, civilian pilots and plane owners could "assist in minimizing aerial sabotage and subversive activities." A War Department–organized "Civil Air Guard" could potentially "perform many missions such as coastal and other patrol, tow target flying, antiaircraft searchlight and sound detector training, transportation of personnel and supplies of all types, etc."[64] In a reply to a letter from Jones relative to the organization of an "Air Vigilante," Stimson referenced the informal board. He advised that the War Department's study indicated the advisability of organizing civilian airmen and aircraft owners into a "quasi-military organization that may be trained to conduct flight missions of a semi-military nature."[65] Stimson explained that, although the War Department had not yet formulated complete plans, its proposed organization would be taken into consideration before the advancement of any CAA-approved organization of an Air Vigilante.[66]

Out of all the competing plans, the CADS proposal by Wilson, Beck, and Gannett represented the most complete framework for a viable organization and the foundation for further refinement. LaGuardia tasked Landis, who was privy to the War Department board's "Civil Air Guard" plan, to combine its thinking with OCD's CADS proposal. From this effort a proposed "Civil Air Patrol" emerged.[67] On 29 September, LaGuardia wrote to Stimson, Jones, and Secretary of the Navy Frank Knox about OCD's plan for CAP to "make available as efficiently as possible the existing civilian aviation potential for national defense and . . . raise the level of skill of the civilian aviation structure to improve the potential value for national defense." Although the letter lacked particulars on what CAP could accomplish, OCD's director mentioned the potential to organize a "corps d'elite, probably known as the 'Civil Air Reserve,' " of higher-than-average qualified pilots and personnel who might be of "rather unusual value to the national defense effort."[68]

The three departments all approved LaGuardia's proposal and agreed to cooperate.[69] LaGuardia requested Wilson's services from New Jersey governor Edison to work with Landis in ironing out details of CAP's composition.[70] On 7 November, the Army Air Forces convened a special board to determine the basis on which the War Department would enter into OCD's plan for CAP.⁷¹ The board found that "carefully selected civilian pilots, aircraft and engine mechanics, radio operators, and airport attendents [sic], properly organized and trained can, in many instances, supplement and assist the military forces in various activities."⁷² Army officers recommended a cooperative agreement with OCD for CAP training and organization but concluded the responsibility for both rested with OCD. War Department manuals, textbooks, circulars, and other unclassified training publications would be made available to CAP upon request. A retired Army officer would act as CAP's national commander with no expenditure of War Department funds other than for personnel assigned to OCD.⁷³

For a period from late September to early October, the AOPA made a clumsy, self-centered effort to form a partnership between OCD's efforts and their Air Guard. In a proposed OCD-AOPA agreement, OCD would cover AOPA expenses to train only licensed pilots in the Air Guard. The training would include Army familiarization courses, precision flying, aerial photography, formation flying, and shore and ocean patrol and reporting, among other skills. A research group would be formed to "continuously study and devise novel means whereby civil aviation can be used to assist in defense."⁷⁴ Landis coyly implied OCD was exploring other options.⁷⁵ AOPA continued to press the matter until Landis explained, "A plan has been adopted and is actively in course of work."⁷⁶ Wilson was asked if he wanted to do anything with the AOPA proposal or to include the organization in the OCD-CAP mailing list, and he replied with the concise answer: No.⁷⁷

In late November, the Army tapped New York City native Maj Gen John F. Curry to serve as CAP's national commander.⁷⁸ A graduate of the US Military Academy, Curry flew as part of Gen John J. Pershing's Punitive Expedition in 1916. He later served as chief of staff for the Second Army Air Service in France in 1918. From 1931 to 1935, Curry was commandant of the Air Corps Tactical School at Maxwell Field, Alabama, presiding over a faculty that debated and refined new strategic uses of airpower that burst upon the world in the early 1940s. At the time of his selection to helm CAP, he directed the Rocky Mountain Technical Command at Lowry Field, Colorado. A graduate of the Air Corps Tactical School, the Command and General Staff School, and the Army War College, Curry brought to CAP a deep understanding of aviation's potential in warfare, and a rich professional network of senior military leaders.⁷⁹

Figure 3. Maj Gen John F. Curry, first Civil Air Patrol national commander. (Photograph courtesy of the Morse Center.)

With seemingly all but the state wing commanders finalized by late November, LaGuardia ordered OCD to "hold everything up" and make no formal or informal communication about CAP without his personal signature.[80] On 1 December, LaGuardia approved the creation of a CAP division within OCD and placed his signature on a simplified establishment order intended for inclusion in an information booklet about CAP.[81] A week later, caught up in the shock of the Imperial Japanese Navy's attack on American forces in Hawaii, LaGuardia initialed OCD Administrative Order No. 9, formally establishing CAP on 8 December.[82] That evening, he announced CAP's existence in a national radio broadcast; OCD issued a press release defining the new organization as "an organization of the civilian aviation resources of the nation for national defense service."[83]

For the civilian aviation community, CAP's establishment also coincided with restrictions on civil aviation. On 8 December, the day after the Japanese attack in Hawaii, the CAA implemented contingency plans developed in 1940. The CAA suspended all pilot certificates except for those pilots employed by scheduled air carriers under contract with the federal government. Hinckley asked governors to assign police to guard airfields and private aircraft.[84] Two days later, the CAA issued guidance that aircraft could be flown with the reinstatement of the pilot certificates, provided pilots submitted evidence of their American citizenship and loyalty to the United States government.[85]

On 16 December, the CAA published a series of orders governing civilian aviation for the foreseeable future. All civilian aircraft not stored or staked out under 24-hour guard had to be rendered incapable of operation by removal of essential parts until passage of the national emergency. Airport managers received new guidance to approve or disapprove all local flights within 10 miles of the airport and to log the arrival and departure of all aircraft. After 8 January 1942, all civilian pilots would have to carry a new CAA pilot identification card containing fingerprints, a photograph, and signature in addition to a current pilot certificate. Failure to possess and carry the card would result in the grounding of the pilot.[86]

Conveniently, CAP's membership requirements meshed with CAA's new requirements for private flying. Members had to be at least 16 years and older to enroll (only 18+ years for flight duty) for the duration of the war, provided they were native-born or naturalized citizens of good character, submitted a FBI fingerprint card and photograph, and passed a background check.[87] CAP had a policy of absolutely no discrimination as to race, creed, color, or sex. As emphasized by Curry, "each member is to be accepted and assigned to duties strictly upon the basis of his or her experience and record of performance."[88] Joining CAP and becoming a member ensured that anyone with a desire to serve their country and potentially fly could have the opportunity. The overlap of CAP's membership requirements—notably the photograph and fingerprint card—with those of the CAA eventually raised eyebrows with the Bureau of the Budget due to the duplication of effort but was resolved in the new year.[89]

Other aspects of CAP, including equipment for operations and training, from aircraft to facilities, were provided by the members themselves. CAP would not provide flight instruction. Members would either be trained, licensed pilots or would obtain flight training on their own time and dime. Prospective members were not exempt from Selective Service and were informed that they would not be used by CAP (or OCD) for combat duty. Members had to provide their own uniforms with distinctive insignia and take an oath of allegiance to the president. Volunteers were organized into flights, squadrons, and groups within state-based wings. The wings, commanded by a governor-appointed volunteer, reported to regional commanders and they in turn to CAP's national commander. Wings maintained cooperative relations with state civilian defense offices and aviation commissions. Regional CAP commanders would function as intermediaries between the national headquarters and the wings.[90]

Accompanying the public announcement of the new organization, OCD released a booklet authored by Wilson with background information about CAP, together with membership applications, nationwide. The booklet, *Civil Air Patrol: Organization, Purpose, Program, Enlistment*, included three pages of questions and answers. Under the question "what specific services could a well organized and trained CAP render to the national defense?" were 11 suggestions. These included guarding airports, providing courier services, conducting observation patrols of back-country or long stretches of uninhabited coastal areas, towing targets, ferrying aircraft, serving as spotters for the Ground Observer Corps, or searching for crashed military aircraft. Noting the list as "merely suggestive," the booklet added, "the extent of the emergency will determine the variety of uses requested by the Armed Services." The future would carry its own answer.[91]

The remainder of December 1941 proved a blur of activity for CAP's small administrative office setup in rooms 1009 and 1011 in the Dupont Building in Washington, DC. On 9 December, LaGuardia named the Aviation Planning Staff authorized under Administrative Order No. 9. He tasked Landis, his aviation aide, to chair the staff which would research and advise OCD's director on any suggested programs or operations for CAP. Curry would direct the operation of CAP, subject to policy and program directives from the Aviation Planning Staff. LaGuardia named Wilson as acting executive officer to serve under Curry, organizing and directing CAP National Headquarters.[92] From 11 to 13 December, Curry met with the governor-appointed wing commanders in Washington to explain the mission of organizing a CAP wing and subordinate groups, squadrons, and flights in their respective states. Under Wilson's direction, CAP National Headquarters mailed out membership applications, FBI fingerprint cards, and 125,000 booklets across the nation.[93] In late December with CAP organizational development well underway, Landis believed CAP would have operational units in one-third of the nation's counties by 15 January 1942.[94] The first challenge, establishing a presence for civil aviation in home defense, had been met. Now the hard work to build, train, and field the organization would be tested—by both the nation's military and a foreign enemy.

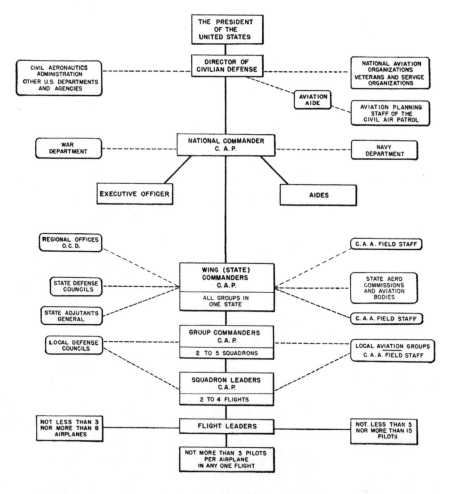

Figure 4. Civil Air Patrol organizational chart as of December 1941. (Image courtesy of the National Archives and Records Administration, College Park, Maryland.)

Notes

Epigraph. Fiorello H. LaGuardia to Frank Knox and Jesse H. Jones, 29 September 1941, Folder "Confidential CAP Chronological," Box 1, Entry 12, General Correspondence, 1941–May 1942, Chronos–020, Record Group 171, Records of the Office of Civilian Defense, National Headquarters (RG171), National Archives and Records Administration, College Park, MD (NARA).

1. Wilson, *I Walked with Giants*, 11–93.
2. Gordon, *The Lafayette Flying Corps*, 473.
3. Wilson, *I Walked with Giants*, 94–146; and Sloan, *Wings of Honor*, 376–77, 386.
4. Wilson, *I Walked with Giants*, 176.
5. Wilson, 178–81.
6. Wilson, 191–217.
7. Wilson, 272.
8. Wilson, 272.
9. Wilson, 267–79; and Gill Robb Wilson to Audley H. F. Stephan, 21 August 1940, Folder "507 Civil Air Patrol," Box 111, Entry 10, National Headquarters, General Correspondence, 1940–1942, 502 to 511, RG171, NARA. In his autobiography, Wilson reports the Red Cross "weren't interested" in his proposal. Wilson, *I Walked with Giants*, 279.
10. Hill, *The Minute Man in Peace and War*, 523–32; and Wilson, *I Walked with Giants*, 193, 203–11.
11. "Milton Knight, 65, Glass Executive," *New York Times*, 11 November 1971, 50; "Toledo Starts Air Amateurs' Reserve Corps," *Dayton Daily News*, 26 December 1938, 13; Grebe, "Toledo's C.A.R.," 35–36; and Civilian Air Reserve (CAR), *Organization Handbook*, 10.
12. CAR, *Organization Handbook*, 4.
13. CAR, 4, 10.
14. CAR, 9.
15. CAR, 3–12.
16. "Civil Air Reserve Program Gains Momentum by Local Effort," *National Aeronautics*, October 1940, 21; "Developments on Civil Air Reserve," *National Aeronautics*, December 1940, 14, 43; "Florida Organizes Civil Air Reserve Unit," *National Aeronautics*, December 1940, 20; "Denver CAR Stages 'Mercy' Flight," *National Aeronautics*, September 1941, 25; "Civil Air Defense," *National Aeronautics*, October 1941, 12–14; and "Program of Nation-wide Expansion Now Planned by Civil Air Reserve: Units will Study Aviation Subjects," *State Journal* (Lansing, MI), 28 March 1940, 1.
17. Earle L. Johnson to John E. Shennett, memorandum, subject: Civilian Air Reserve, 14 October 1942, BLS, CAP-NAHC.
18. Yoshpe, *Our Missing Shield*, 59–61.
19. Jordan, *U.S. Civil Defense Before 1950*, 34–36.
20. National Defense Advisory Commission letter and memorandum, "State and Local Cooperation in National Defense," reproduced in chapter titled "State and Local Cooperation in the National Defense Program" in Council of State Governments, *The Book of the States*, 4:33–42.
21. Harries, *Last Days of Innocence*, 49–60; Millett and Maslowski, *For the Common Defense*, 338–42; and Finnegan, *Against the Specter of a Dragon*.
22. Dallek, *Defenseless Under the Night*, 105–6.

23. Department of Commerce, Civil Aeronautics Administration, press release for 4 October 1940, Folder "811 Advisory Board," Box 345, Entry UD 1, Civil Aeronautics Administration, Central Files, 807–811P, Federal Aviation Administration, RG237, NARA; Milton Knight to All Wing Commanders and Group Commanders regarding National Convention, 19 July 1940, Folder 1, "Personal Correspondence, August 1939–December 1942," Box 7, ELJ, WRHS; Mauck, "Civilian Defense in the United States," chap. 9, 2; "Milton Knight to Attend First Meet of Air Board," *Toledo Blade*, 2 October 1940, 13; and "Developments on Civil Air Reserve," *National Aeronautics*, December 1940, 14, 43.

24. Gill Robb Wilson to Audley H. F. Stephan, 21 August 1940, Folder "507 Civil Air Patrol," Box 111, Entry 10, RG171, NARA.

25. Gill Robb Wilson to Audley H. F. Stephan, 21 August 1940.

26. Gill Robb Wilson to Audley H. F. Stephan, 21 August 1940.

27. Audley H. F. Stephan to Frank Bane, 27 August 1940; and Frank Bane to Audley H. F. Stephan, 29 August 1940, Folder "507 Civil Air Patrol," Box 111, Entry 10, RG171, NARA.

28. Hill, *Minute Man in Peace and War*, 528; and Gross, *Prelude to the Total Force*, 2.

29. Headquarters Army Air Forces, Routing and Record Sheet, Subject "Air Vigilante" or "Civil Air Guard," 15 August 1941, Folder "SAS 381 Civil Defense," Box 111, Henry Harley Arnold Papers (HHA), Manuscript Division, Library of Congress, Washington, DC (LOC); Mauck, "Civilian Defense in the United States, 1941–1945," chap. 9, 2; "AOPA News," *Flying and Popular Aviation* 27, no. 2 (August 1940): 49–50; "AOPA News," *Flying and Popular Aviation* 27, no. 3 (September 1940): 49–50; and "AOPA News," *Flying and Popular Aviation* 28, no. 3 (March 1941): 53.

30. Dallek, *Defenseless Under the Night*, 111–14.

31. Thomas H. Beck to Franklin D. Roosevelt, 22 April 1941, attached to letter of Thomas H. Beck to Kendall K. Hoyt, 16 September 1942, Kendall King Hoyt Papers (KKH), CAP-NAHC.

32. *Cong. Rec.*, 77th Cong., 1st sess., 1941, vol. 87, pt. 4: 3687. The bills include: H.R. 4664 (introduced 6 May 1941); H.R. 4670 (introduced 7 May 1941); H.R. 4746 (introduced 14 May 1941); H.R. 4758 (introduced 15 May 1941); and S. 1554 (introduced 23 May 1941).

33. "To Establish a Civilian Air Reserve" (Civilian Air Reserve Act of 1941), HR 4664, 77th Cong., 1st sess., introduced 6 May 1941.

34. Donald H. Connolly to Jesse H. Jones, memorandum, subject: H.R. 4664—a bill "to establish a Civilian Air Reserve to be organized and trained by the War and Navy Departments, and for other purposes," 31 May 1941, Folder "811 Advisory Board," Box 345, Entry UD 1, Civil Aeronautics Administration, Central Files, 807–811P, Federal Aviation Administration, RG237, NARA.

35. Franklin D. Roosevelt, Executive Order 8757, "Establishing the Office of Civilian Defense in the Office for Emergency Management of the Executive Office of the President," *Federal Register* 6, no. 100 (22 May 1941): 2517–18.

36. Guy P. Gannett to Kendall K. Hoyt, 2 October 1942; and Kendall K. Hoyt, "Civil Aviation Services Chronology of Origin and Progress," 22 January 1942, attached to letter of Thomas H. Beck to Kendall K. Hoyt, 16 September 1942, KKH, CAP-NAHC.

37. Press release from Washington Correspondent, Washington, DC, for *Portland Press Herald*, "La Guadia [sic] Names Civil Air Reserve; Acts on Maine Man's Plan to Aid Defense: Guy P. Gannett to Serve with Gil [sic] Robb Wilson, Thomas H.

Beck," 13 June 1941, KKH, CAP-NAHC; and "LaGuardia Names Air Defense Group," *Miami News* (FL), 15 June 1941, 6.

38. Fiorello H. LaGuardia to Gill Robb Wilson, 12 June 1941, BLS, CAP-NAHC.
39. Mauck, "Civilian Defense in the United States," chap. 9, 1.
40. Gill Robb Wilson to Charles Edison, memorandum, subject: Formation of a Civil Air Guard, 24 March 1941, courtesy of Jill D. Paulson, Libertyville, IL.
41. Gill Robb Wilson to Charles Edison, memorandum, subject: Air Reserve, 27 March 1941, courtesy of Jill Robb Paulson.
42. "Edison Urges Formation of Civil Air Guard," *Courier-News* (Bridgewater, NJ), 8 April 1941, 2.
43. "Civil Air Guard Planned for N.J.," *Daily Journal* (Vineland, NJ), 22 May 1941, 1.
44. "Jersey to Form State Air Guard," *Asbury Park Press* (Asbury, NJ), 14 June 1941, 9.
45. "Civilian Pilot Recruiting to Start in State July 10," *Asbury Park Press*, 3 July 1941, 1, 10; Larry Woodruff, "Plane Paragraphs: Pilots and Aircraft Owners at Hadley Cooperate with New Jersey Wing, Civil Air Defense Services," *Central New Jersey Home News* (New Brunswick, NJ), 25 July 1941, 8; "N.J. Air Defenses Model for States," *Courier-Post* (Camden, NJ), 22 November 1941, 4; and New Jersey Defense Council, *New Jersey Wing: Civil Air Defense Services*.
46. Kendall K. Hoyt, "Civil Aviation Services Chronology of Origin and Progress," 22 January 1942, attached to letter of Thomas H. Beck to Kendall K. Hoyt, 16 September 1942, KKH.
47. Gill Robb Wilson to Fiorello H. LaGuardia, 24 June 1941, as transmittal letter for "Proposed Plan for Organization of Civil Air Defense Resources," CAP-NAHC.
48. Office of Civilian Defense (OCD), "Proposed Plan for Organization of Civil Air Defense Resources," 24 June 1941, 1–2, CAP-NAHC. See appendix G.
49. OCD, "Proposed Plan for Organization of Civil Air Defense Resources," 24 June 1941, 3–6, CAP-NAHC.
50. Douhet, *Command of the Air*, 83.
51. Douhet, 82–87; and Meilinger, *Paths of Heaven*, 16–17.
52. Mitchell, *Our Air Force: The Keystone of National Defense*, xxiv–xxv, 194–95, 202–3; Mitchell, *Winged Defense: The Development and Possibilities of Modern Air Power*, 6, 97–98, 112–13; and Meilinger, *The Paths of Heaven: The Evolution of Airpower Theory*), 98–101.
53. Henry L. Stimson to Jesse H. Jones, 20 June 1941, Folder "SAS 381 Civil Defense," Box 111, HHA, LOC.
54. Jesse H. Jones to Henry L. Stimson, 17 July 1941, Folder "SAS 381 Civil Defense," Box 111, HHA, LOC.
55. Daso, *Hap Arnold and the Evolution of American Airpower*, 103–68, 230–31; Huston, *American Airpower Comes of Age*, 4–108; and Meilinger, *Paths of Heaven*, 188–89.
56. Pisano, *To Fill the Skies with Pilots*, 3, 10–12, 30, 58–90; and Strickland, *The Putt-Putt Air Force*, 3–4, 7–13. At the time of its abolishment in 1946, CPTP had helped produced 435,165 pilots for the nation's war effort. Strickland, *Putt-Putt Air Force*, iii.
57. Henry H. Arnold to Fiorello H. LaGuardia, 19 July 1941, Binder "Legal Status, Administrative Concepts, and Relationship of the Civil Air Patrol, 1941 to 1949," CAP-NAHC.

58. Henry H. Arnold to Lorenzo D. Gasser, 28 July 1941, Binder "Legal Status, Administrative Concepts, and Relationship of the Civil Air Patrol, 1941 to 1949," CAP-NAHC.

59. Kendall K. Hoyt, "Civil Aviation Services Chronology of Origin and Progress," 22 January 1942, attached to letter of Thomas H. Beck to Kendall K. Hoyt, 16 September 1942, KKH, CAP-NAHC.

60. Sloan, *Wings of Honor*, 206, 377–79, 387.

61. Hudson, "Reed G. Landis and 'The Last Good War,' " 127–41; "Reed Landis Appointed Aid to La Guardia," *Chicago Tribune*, 30 July 1941, 5; "Landis's Son to Serve," *Cincinnati Enquirer*, 30 July 1941, 18; and Press release, OCD, "For Tuesday A.M.'s, 9 December 1941," Folder labeled "Civil Air Patrol," Box 192, World War II Papers, 1939–1947, State Archives, North Carolina Office of Archives and History, Raleigh.

62. The board consisted of Maj Alexis B. McMullen, A-3, Air Staff; Maj Lucius Ordway, Buildings and Grounds, Office of the Chief of the Air Corps; and Maj Guido R. Perera, A-1, Air Staff.

63. Milton Knight to Earle L. Johnson, 26 August 1941, Folder 1, Box 7, ELJ, WRHS.

64. Headquarters Army Air Forces, Routing and Record Sheet, Subject "Air Vigilante" or "Civil Air Guard," 15 August 1941, Folder "SAS 381 Civil Defense," Box 111, HHA, LOC.

65. Henry L. Stimson to Jesse H. Jones, 9 September 1941, Folder "020 War Dept./Vol. 7 [Folder 2 of 2–January 1942 August 1941]," Box 6, Entry UD1, Central Files 020–War Department (to) August 1, 1941, Civil Aeronautics Administration, RG237, NARA.

66. Henry L. Stimson to Jesse H. Jones, 9 September 1941, Folder "020 War Dept./Vol. 7 [Folder 2 of 2–January 1942 August 1941]," Box 6, Entry UD1, Central Files 020–War Department (to) August 1, 1941, Civil Aeronautics Administration, RG237, NARA.

67. Lucius P. Ordway to Reed Landis, 3 September 1941 with attached "Outline of Form of Organization of Civil Air Guard," Binder "Legal Status, Administrative Concepts, and Relationship of the Civil Air Patrol, 1941 to 1949," CAP-NAHC.

68. Fiorello H. LaGuardia to Frank Knox and Jesse H. Jones, 29 September 1941, Folder "Confidential CAP Chronological," Box 1, Entry 12, General Correspondence, 1941–May 1942, Chronos–020, RG171, NARA. Landis drafted both letters.

69. Neprud, *Flying Minute Men*, 22–23; Fiorello H. LaGuardia to Henry L. Stimson, 8 October 1941, Folder "Confidential CAP Chronological," Box 1, Entry 12, General Correspondence, 1941–May 1942, Chronos–020, RG171, NARA; Mauck, "Civilian Defense in the United States," chap. 9, 3; and Henry L. Stimson to Fiorello H. LaGuardia, 29 October 1941, Reel 44552, AFHRA.

70. Telegram from Fiorello H. LaGuardia to Charles Edison, 7 October 1941, Folder "Confidential CAP Chronological," Box 1, Entry 12, General Correspondence, 1941–May 1942, Chronos–020, RG171, NARA.

71. This board consisted of Brig Gen George E. Stratemeyer (president), Col Harry H. Blee, Maj Alexis B. McMullen, and Maj Lucius P. Ordway.

72. War Department, Headquarters Army Air Forces, Special Orders No. 53, 4 November 1941; George E. Stratemeyer to Chief of the Army Air Forces, memorandum, subject: Civil Air Patrol, 8 November 1941; and Report of Proceedings of Board of Officers, Headquarters Army Air Forces, 8 November 1941, Binder "Legal Status, Administrative Concepts, and Relationship of the Civil Air Patrol, 1941 to 1949," CAP-NAHC.

73. War Department, Headquarters Army Air Forces, Special Orders No. 53, 4 November 1941; George E. Stratemeyer to Chief of the Army Air Forces, memorandum, subject: Civil Air Patrol, 8 November 1941; and Report of Proceedings of Board of Officers, Headquarters Army Air Forces, 8 November 1941, Binder "Legal Status, Administrative Concepts, and Relationship of the Civil Air Patrol, 1941 to 1949," CAP-NAHC.
74. Laurence P. Sharples to Reed Landis, 26 September 1941, attachment, "O.C.D.-A.O.P.A. Agreement, 26 September 1941," BLS, CAP-NAHC.
75. J. B. Hartranft Jr. to Reed G. Landis, 30 September 1941; and J. B. Hartranft Jr. to Reed G. Landis, 10 October 1941, BLS, CAP-NAHC.
76. Reed G. Landis to J. B. Hartranft Jr., 14 October 1941, BLS, CAP-NAHC.
77. Kathryn Godwin to Gill Robb Wilson, 4 November 1941; and L. H. [full name not provided] to Gill Robb Wilson, 15 November 1941, BLS, CAP-NAHC.
78. Fiorello H. LaGuardia to Henry H. Arnold, 21 November 1941, Folder "Confidential CAP Chronological," Box 1, Entry 12, General Correspondence, 1941–May 1942, Chronos-020, RG171, NARA. Evidently Curry's orders did not post until around 3 December 1941, although another source mentions "Simultaneously with the issuance of Administrative Order No. 9 Major General Curry was designated National Commander of the Civil Air Patrol. . . ." Administrative Officer Report, 9 March 1942; Reed G. Landis to John F. Curry, 3 December 1941, BLS, CAP-NAHC; message entry for 3 December 1941, Folder "Military Miscellaneous—Mail Lists August–December 1941," Box 207, HHA, LOC; and Fiorello H. LaGuardia to J. Melville Broughton, 28 November 1941, Folder labeled "Civil Air Patrol," Box 192, World War II Papers, 1939–1947, State Archives, North Carolina Office of Archives and History, Raleigh.
Prior to Curry's selection, LaGuardia proposed retired Maj Gen James E. Fechet, former Chief of the Air Corps, and the Army Air Forces board of 4 November 1941 agreed with the selection of Fechet, "if possible," to act as CAP's national commander. Due to difficulties in bringing the retired Fechet back to active duty, Reed Landis floated the idea of extending the service of Maj Gen Frank P. Lahm to allow him to serve as CAP National Commander. The subsequent selection of Curry appears to have resulted from a private conversation between Landis and Arnold in late November. Fiorello H. LaGuardia to Henry L. Stimson, 8 October 1941; Report of Proceedings of Board of Officers, Headquarters Army Air Forces, 8 November 1941, Binder "Legal Status, Administrative Concepts, and Relationship of the Civil Air Patrol, 1941 to 1949," CAP-NAHC; Reed G. Landis to Fiorello H. LaGuardia, memorandum, 5 November 1941; and Reed G. Landis to Henry H. Arnold, 19 November 1941, Reel 44552, AFHRA.
79. Biographies, "Major General John Francis Curry," US Air Force, https://www.af.mil/About-Us/Biographies/Display/Article/2286532/major-general-john-francis-curry/; Meilinger, *Bomber*, 15–16; and Finney, *History of the Air Corps Tactical School*, 23–38, 76, 83, 102–4.
80. Fiorello H. LaGuardia to T. Semmes Walmsley, telegram, 25 November 1941, Folder "507 Civil Air Patrol," Box 111, Entry 10, General Correspondence, 1940–1942, 502 to 511, RG171, NARA.
81. OCD, *Civil Air Patrol*, 3; and Cecile Hamilton to Gill Robb Wilson, memorandum, subject: Activities on Tuesday, 2 December 1941, Reel 44552, AFHRA. CAP celebrates 1 December 1941 as its official birthday, but the legal, administrative basis for CAP's existence is OCD Administrative Order No. 9, initialed by LaGuardia and issued 8 December 1941. The simplified establishment order dated 1 December 1941 bearing LaGuardia's signature is explained by Reed Landis in a letter to LaGuardia of

15 November. Landis noted how the OCD staff was "afraid that a long, detailed order may first confuse the people in the field and secondly require so much revision as the show develops as to cause further confusion among those who will have seen the initial booklet but not understand completely the interior administration reasons requiring changes and ways of doing various things." Reed G. Landis to Fiorello H. LaGuardia, 15 November 1941, Reel 44552, AFHRA.

To further add to the confusion about CAP's true establishment date, in CAP National Headquarters General Orders No. 1 issued on 7 January 1942, the document declares that on 1 December 1941, "the following statement was issued from the Office of Civilian Defense" listing CAP as being "established"—whereas the same document, on the same page, reports Administrative Order No. 9 as "establishing the Civil Air Patrol." This order, ironically, was partially drafted by both Reed Landis and Wilson. John F. Curry, OCD, Civil Air Patrol National Headquarters, General Orders No. 1, 7 January 1942, Reel 38909; and Reed G. Landis to Gill Robb Wilson, memorandum, subject: Civil Air Patrol, 9 December 1941, Reel 44552, AFHRA. For a broader discussion of the confusion over the date of CAP's establishment, see Blazich, Civil Air Patrol National History Program, "'Founding' versus 'Establishment.'"

82. OCD, Administrative Order No. 9, 8 December 1941, Folder "8 December 1941 A09; Subject: Establishing Civil Air Patrol," Box 39, Entry 10, Office Procedures & Personnel Memos, Reports & Awards Office, July 1941–May 1945, Administrative Orders, RG171, NARA.

83. Press release, OCD, "For Tuesday A.M.'s, 9 December 1941," Folder labeled "Civil Air Patrol," Box 192, World War II Papers, 1939-1947, State Archives, North Carolina Office of Archives and History, Raleigh; and OCD, *Civil Air Patrol*, 11. CAP's official definition of April 1942 is listed as "a volunteer civilian defense organization established for the purpose of mobilizing and training the civil aviation personnel and equipment of the nation not otherwise actively engaged full-time in governmental service or commercial air transport activities, that such may be available and effective for auxiliary service to the armed and civil defense forces of the United States." Kathryn Godwin to Mr. Straw, memorandum, subject: Definition of Civil Air Patrol, 8 April 1942, BLS, CAP-NAHC.

84. Wilson, *Turbulence Aloft*, 84, 86; "All Private Planes are Grounded by CAA; Pilot Licenses, Except on Lines, Suspended," *Washington Post*, 8 December 1941, 3; "Only Airliners Now May Fly, C.A.A. Orders," *Washington Post*, 8 December 1941, 20; "Governors Asked to Guard Airports," *Washington Post*, 9 December 1941, 9; and George A. Vest to J. Melville Broughton, telegram, 9 December 1941, Folder "Civilian Defense Activities–Air Patrol; Fire School; Warning Posts, etc.," Box 29, Agencies, Commissions, Departments, and Institutions, 1941–1944, Government J. Melville Broughton Papers, North Carolina State Archives, Raleigh, NC.

85. Wilson, *Turbulence Aloft*, 86–87; and George W. Vest to J. Melville Broughton, 18 December 1941, with attachments, Folder "Civil Air Patrol," Box 192, OCD, World War II Papers, 1939-1947, Military Collection, North Carolina State Archives, Raleigh, NC.

86. Wilson, *Turbulence Aloft*, 86–87; Department of Commerce, Civil Aeronautics Administration, Second Region, George W. Vest to All State Aviation Directors, All Airport Managers, memorandum, 16 December 1941; and George W. Vest to J. Melville Broughton, 18 December 1941, with attachments, Folder "Civil Air Patrol," Box 192, OCD, World War II Papers, 1939-1947, Military Collection, North Carolina State Archives, Raleigh, NC.

87. Naturalized US residents required a minimum of 10 years of citizenship and have to have been born in a country that was not an enemy nation. John F. Curry to

all wing commanders, memorandum, subject: Citizenship Requirements for Civil Air Patrol, 2 March 1942, Folder "Civil Air Patrol 15 July 1942," Box 21, Entry 16A, General Correspondence, 1941–May 1945, Civil Air Patrol, RG171, NARA.

88. OCD, press release, 28 January 1942, Binder "Legal Status, Administrative Concepts, and Relationship of the Civil Air Patrol, 1941 to 1949," CAP-NAHC.

89. Herman Kehrli to Bernard L. Gladieux, memorandum, subject: Civil Air Patrol, 6 March 1942; Ivan Hinderaker to Bernard L. Gladieux, memorandum, subject: Interview with Mr. Stanley Tracy, Chief of the Fingerprint Section, FBI, 9 March 1942; Ivan Hinderaker to Bernard L. Gladieux, memorandum, subject: Civil Aeronautics Authority and Civil Air Patrol Pilot and Aircraft Roster Facilities, 19 March 1942; and Bernard L. Gladieux to L. C. Martin, memorandum, subject: The CAP and CAA Rosters, 27 March 1942, Folder "OCD Civil Air Patrol," Box 10, Entry 107A, Budgetary Administration Records for Emergency and War Agencies and Defense Activities, 1939–1949 (file 31.19), Record Group 51, Records of the Office of Management and Budget (RG51), NARA.

90. OCD, *Civil Air Patrol*, 4, 11–13; John F. Curry, OCD, CAPNHQ, to all CAP unit commanders, memorandum, subject: Procedure concerning Federal Bureau of Investigation check of Civil Air Patrol applicants, 2 March 1942, Folder "Civil Air Patrol 15 July 1942," Box 21, Entry 16A, General Correspondence, 1941–May 1945, Civil Air Patrol, RG171, NARA; and John F. Curry, OCD, CAPNHQ, General Orders No. 1, 7 January 1942, Reel 38909, AFHRA.

CAP regional commanders were disestablished in August 1942 as OCD was unable to secure nine Army Air Forces officers to staff the positions. Thereafter, wing commanders reported directly to CAP National Headquarters. James M. Landis to Regional Directors and Washington Staff, memorandum, subject: Civil Air Patrol Organization, 14 August 1942, Folder "Civil Air Patrol August 1 Thru," Box 21, Entry 16A, General Correspondence, 1941–May 1945, Civil Air Patrol, RG171, NARA.

91. OCD, *Civil Air Patrol*, 11–13. The booklet first went to the Government Printing Office on 3 December 1941. Among the last-minute changes, Landis removed a question about applicant race from the enlistment form in accordance with Executive Order 8802, which prohibited discrimination in the employment of workers in defense industries or government. Cecile Hamilton to Gill Robb Wilson, memorandum, subject: Activities on Tuesday, 2 December 1941, Reel 44552, AFHRA.

92. Fiorello H. LaGuardia to John F. Curry, 9 December 1941, Folder "507 Civil Air Patrol," Box 111, Entry 10, RG171, NARA. The Aviation Planning Staff consisted of Brig Gen George E. Stratemeyer and Maj Alexis B. McMullen of the War Department, Capt A. W. Bradford and Lt Cmdr A. W. Wheelock of the Navy Department, Earl Southee and C. I. Stanton of the CAA, T. Semmes Walmsley and Reed G. Landis of OCD, and five civil aviation men at large: Roger W. Kahn, Henry King, Harry Coffey, Harry Playford, and Egbert P. Lott.

93. Administrative Office Report, 9 March 1942, BLS, CAP-NAHC.

94. Reed G. Landis to Alexis B. McMullen, 26 December 1941, Binder "Legal Status, Administrative Concepts, and Relationship of the Civil Air Patrol, 1941 to 1949," CAP-NAHC.

Chapter 3

Paukenschlag and the CAP Experiment

The attack was a complete success. The U-boats found that conditions there were almost exactly those of normal peacetime. . . . Although five weeks had passed since the declaration of war, very few anti-submarine measures appeared to have been introduced.

—Vice Adm Karl Doenitz, commander of Germany's U-boats

On 12 January 1942, five German U-boats started sinking merchant shipping off the Eastern Seaboard of North America. The German offensive, codenamed *Paukenschlag* ("Drumbeat"), delivered a jarring blow to the American home front and found the nation's military underequipped and ill prepared to combat the U-boat menace. The Navy's antisubmarine resources at the onset of the war in December 1941 were extremely limited. Rear Adm Adolphus Andrews, commander of the Eastern Sea Frontier comprising almost 1,500 miles of coastline from the Canadian border with Maine to the southern boundary of Duval County, Florida, could marshal a meager force: 20 under-armed, undermanned ships of varying reliability and a motley assortment of 103 aircraft, three-quarters of which were unsuited for either coastal patrol or antisubmarine defense.[1] The frontier war diary admitted the basic fact at hand: "When we entered the war, we did not have the naval strength required to defend the merchant shipping we needed."[2]

At the request of the Navy, Lt Gen Hugh A. Drum, head of the Army's Eastern Defense Command, had ordered the I Bomber Command, a training unit, to start antisubmarine patrols on 8 December 1941. Despite frantic efforts to augment the command with aircraft from the First Air Force, the Army initially mustered only 100 twin–engine aircraft to patrol the entire Eastern Seaboard.[3] Lack of coordination further hampered American antisubmarine efforts, as the Navy and Army forces operated under parallel command structures and had no provisions for coordination or information sharing. Naval forces reported to Eastern Sea Frontier, while the Army's I Bomber Command under Brig Gen Arnold N. Krogstad received its directives from the Eastern Defense Command.[4]

From late January into February 1942, Andrews was working tirelessly to acquire more forces. The first five Drumbeaters sank 23 merchant ships totaling 151,505 tons off the East Coast. Under the direction of Commander-in-Chief, US Fleet (COMINCH) Adm Ernest J. King, on 24 January Andrews received operational control of 44 PBY Catalina flying boats assigned to the Atlantic Fleet for antisubmarine patrols.[5] The Coast Guard provided Andrews with 40 OS2U-3 Kingfisher observation aircraft, patrolling from stations along the East Coast. King also gave Andrews temporary use of 10 destroyers, and the sea frontier commander received word in mid-February that 24 British converted fishing trawlers and 10 corvettes would be sent across the Atlantic on loan for patrol and escort duty. Unfortunately, the British vessels would not arrive until March.[6] I Bomber Command, operating independently of Andrews, mustered a mere 119 aircraft, with only 46 in commission, including 9 B-17 heavy bombers capable of long-range patrol and the remainder a mix of B-18s and B-25s medium bombers for patrol of the sea lanes, with crews untrained and ill equipped for antisubmarine warfare.[7]

As the British had learned in World War I, the best defense against U-boat attacks on shipping was to have warships escort merchant vessels into organized convoys. Admiral King recognized the importance of husbanding the Navy's remaining resources for a two-ocean war. Having already suffered heavy losses to the Asiatic and Pacific Fleets, he positioned naval forces to best serve the nation's strategic plans. He decided that all troopships would sail in heavily armed, escorted convoys, thereby placing the lives of Soldiers and Marines above that of bulk cargo. A convoy of merchant ships escorted by destroyers able to defend and attack represented the ideal situation to King. But an unescorted convoy risked concentrating vulnerable targets in a small area and thus was worse than nothing at all. Lost in all this was the difference between a coastal convoy, escorted by nearby land and potentially air cover, and an oceanic convoy. British and Canadian experience had already demonstrated that the latter could survive with light escort protection. Despite British pressure and proven evidence of the efficacy of coastal convoys, King waited for escorts while Andrews and Krogstad searched for interim solutions.[8]

While the Army and Navy cobbled together military resources, CAP leadership began to advocate for its employment to counter the U-boat threat. On 6 January 1942, Curry received a letter from Maj Frank Flynn that drew from his experience in World War I flying coastal patrols off Felixstowe, England; Flynn believed CAP could do something similar along the Pacific Coast to thwart Japanese submarines.[9] Weeks later, Hollywood film producer and private pilot Henry King, a member of LaGuardia's Aviation Planning Staff, suggested the idea of using slow, low-flying aircraft to spot small, suspicious objects or anything resembling a periscope in the water and radio in such sightings to authorities for further investigation.[10] Pennsylvania Wing member Anthony Schmittinger suggested in his letter to Curry that CAP aircraft armed with 150-pound bombs "could make submarine warfare in close to shore tough for Hitler."[11]

Another letter of 24 January from Irving H. Taylor of the Aeronautical Chamber of Commerce of America provided Curry with a detailed report on the British use of light planes for offshore patrolling and spotting of sea mines. In closing, Taylor concluded, "Beyond a question of a doubt the coastal patrol is the most urgently needed defense function which C.A.P. could handle. C.A.P. planes on such patrol work would act as 'beaters' and on occasion would 'flush' enemy subs from hiding and force them out where the combat ships of our Navy and Air Force could put them out of commission. If nothing else, these little planes could keep enemy subs down in the coastal shipping area and thus serve a similar purpose to anti-aircraft which keeps enemy bombers above the 20,000 feet level thus minimizing the effectiveness of the attack."[12] Curry's reply is lost, but Taylor's suggested mission arrived at a pertinent moment in CAP's development. In a 23 January report to Arnold, Curry reported 13,500 membership applications received, accounting for 14.7 percent of total registered pilots nationwide. Existing organizational difficulties notwithstanding, Curry lauded the potential value of CAP, remarking, "It is believed we can organize and conduct within a short time close in (10 miles) shore patrol all around the key coastal spots with great frequency per day with radio equipped ships, thus aiding materially in providing safe waters for coastwise shipping."[13] A week later, Curry again wrote to Arnold about possibilities for the use of CAP aircrews to supplement Army Air Forces operations—on the Pacific Coast, suggesting among other ideas the organization of a close inshore patrol extending 15 miles off the coast to cover the major shipping lanes.[14]

December 1941 to January 1942 saw CAP built and pushed out of the hangar. February 1942 found the organization begin to taxi and look for a mission to test its mettle. At the start of the month Curry could report 275 groups and 687 squadrons in existence, some actively engaged in local missions and others champing at the bit.[15] On 4 February, Arnold asked Brig Gen Donald H. Connolly, administrator of civil aeronautics in the Department of Commerce, for suggestions on the employment of CAP's volunteers and light aircraft.[16] Days later Curry wrote to Drum at the Eastern Defense Command and provided data on CAP and examples of its varied work. Curry noted CAP's "pilots vary from those who are in the air because God lets them stay there to men of extraordinary ability" but added "all of them can be used in some way or another."[17] Drum agreed with Curry on utilizing CAP's members, albeit once a determination could be made as to CAP's place within "the general picture of National Defense."[18]

By late January–early February 1942, CAP's desire for a mission coincided with a need for additional funding to complete wing organization. On 28 January, James M. Landis, successor to LaGuardia as OCD director (and no relation to Reed G. Landis), submitted a request to the Bureau of the Budget for a supplemental appropriation of $114,142 for CAP, $97,000 of which for administrative assistance to wing commanders.[19] James Landis's actions came in response to a funding request from Curry. The general explained how the funds would specifically assist wing commanders by providing administrative assistance to further facilitate pilot registration and accelerate organizing and training CAP units and personnel so as to conduct missions for national, state, and local agencies engaged in the war effort. Many wing commanders the past months had spent considerable sums out of pocket in telephone calls, stenographer services, and other business resources. Curry considered the request small enough to be accepted without question, but either way a decision was needed whether to obtain the funds to continue CAP or "make an immediate decision to cease its operation."[20] Acknowledging his limited ability to outline and forecast what CAP would be doing in the next three months, Landis emphasized to the Budget Bureau that if OCD had a CAP that was "an inefficient organization" with "an improperly conducted program and slipshod operations of requested missions, we will lose the confidence of the war agencies and the opportunity to make civil aviation's contribution to the progress of the war."[21]

Understandably the Budget Bureau sought clarification. Curry attempted to strengthen James Landis's argument by acquiring a statement of the Army's intention to use CAP for assorted missions, signed by Arnold.[22] Arnold, however, felt it inappropriate to intervene.[23] When the Bureau questioned Reed Landis about the functions of CAP on 10 February, he listed several jobs he believed "were of a character to justify the CAP organization for mission duty." One of his five examples included "submarine coastal patrol by a CAP unit." In their discussion with Reed Landis, Budget Bureau representatives believed it necessary for CAP to present "some formal evidence" from the Army to indicate its support and desire to use CAP. Landis indicated he would secure such formal evidence of approval from Arnold. Aware that some in the Army staff viewed CAP unfavorably, Landis believed that if CAP were given a chance to prove its value, it would overcome the naysayers.[24]

Reed Landis apparently made good on his word. Arnold opted to participate through an intermediary. The Air Force chief tasked Lt Col Gordon P. Saville, director of the Army Air Forces' Air Defense Warning System, to work with CAP and report directly to him on CAP's functions in connection with Army Air Forces activities.[25] Reed Landis spoke with Saville about CAP's budget battle and sent him Curry's draft statement of 27 January.[26] When interviewed by the Bureau of the Budget (albeit on matters other than coastal patrol experiment), Saville discussed how CAP and the Army were then engaged in a vicious circle: CAP would not organize until the Army delegated functions, but the Army would not specify the functions until CAP organized. Saville felt, somewhat cynically, that CAP's enlistments comprised two camps: patriots—fliers interested in defending the United States—and amateur pilots using CAP to circumvent CAA restrictions to pursue their hobby in wartime, who just wanted to fly out of "mere self-preservation." An expert on air defense, he saw CAP as something of a "damned menace" to control but also that could be "very useful provided its organization hew strictly to Army specifications."[27]

Reporting directly to Arnold, Saville perhaps echoed some of the air chief's concerns about crafting a working relationship between CAP and the Army. Saville saw a functioning system with both an administrative organization and a control organization. CAP would represent the administrative body with the wings and subordinate units while the Army would handle the control aspects, responsible for issuing orders for missions to be performed. To provide time for both organizations to establish policy and respective relationships, Saville supported the Army providing CAP funding for at least six months to give the organization time to mature. With time, he believed, the Army would probably establish standards for CAP membership.[28]

As Arnold's invisible hand, Saville balanced Army skepticism of CAP with wartime potential. As noted by historian Mae Link, Army personnel, "because of an innate doubt that civilians suddenly could be fitted effectively into a carefully planned military system," were distrustful of civilian capabilities and dependability.[29] CAP's first members, enthusiastic and impatient to do their part for the war effort, failed to recognize and respect the importance of established military procedures. In early 1942, the interactions between CAP and Army authorities at local levels bred perceptions of competition between both organizations for personnel, missions, or funding.

From his position, Arnold saw the value of close liaison between CAP and the Army Air Forces, either in supplementing military activities or relieving the Army of certain assignments. Within the War Department leadership, a consensus existed that once CAP became semimilitarized it could function smoothly with the Army. The military's skepticism could be overcome if the civilian volunteers had the opportunity to prove their effectiveness. What was missing was confidence and trust.[30]

Arguably the vision outlined by Saville tracked with the vision of LaGuardia and his aides who desired OCD to become something akin to a fourth military branch.[31] At war in Europe and the Pacific, the Army had little time for "amateur hour" from the CAP membership. In a 12 February letter to California wing commander Bertrand Rhine, Curry made mention of forthcoming tentative tables of organization, rank structure, commissions, and warrants. "As you can see," wrote Curry, "we are tending more and more toward a real semimilitary organization. Everyone seems to want that and the Air Staff is insistent that we get some discipline before we can be used."[32] When released on 10 March, the tables of organization for wings, groups, and squadrons established a simple numerical relative rank and grade for personnel with duty title as a title of address. Wing commanders received a relative rank and grade of 1; a group commander, 2; a wing staff officer, 3; and pilots and observers, 6. In simpler terms, the system mirrored the Army in all but commission and titles of ranks.[33]

In early March 1942, Reed Landis articulated his vision for the future of CAP. Reflecting on the matter in a memorandum to LaGuardia's successor at OCD, he addressed the civil-military divide and outlined his vision for the future of CAP.[34] He believed the immediate military needs of CAP's service could be met by organizing a small element of existing civil aviation into a "distinctly military body" and by military discipline. At the same time, he feared the death of those aspects of civil aviation not involved with CAP. To maintain and strengthen civil aviation for the postwar world, Landis believed civilian flight should be made contingent on CAP membership, essentially militarizing the entirety of the nation's private pilots under OCD. Noting how "CAP was originally located in OCD because no one else wanted it!," he thought CAP would be best served if moved to the Commerce Department—if moved at all—and that OCD should take great pride in the conception and growth of "such a healthy member of our war family."[35]

Days later, Reed Landis wrote another, more consequential memorandum, this time to Curry. Regarding the adoption of any plan to harness civilian aviation for the war effort, he emphasized the utility of the substantial number of CAP's pilots who were ineligible for a commission in either the Army or Navy, much less membership in the Army's Specialist Reserve.[36] Aware, however, that as the war effort increased the requirements for commissions and qualifications for the Specialist Reserve would decline, Landis thought it advisable for the War Department to "take over Civil Air Patrol and establish it as the Army Air Force Auxiliary with full disciplinary control over the entire membership."[37] The Army could then assign CAP all the personnel and functions then contemplated for the aviation section of the Specialist Reserve.

If the Army chose this course of action, Landis thought it best to merge the weaker wings with the stronger ones to produce an overall more militarized, capable force. CAP, if under the direct command of the Army Air Forces, could help release Army personnel and equipment from coastal patrol and other operations.[38] Between OCD and the War Department, the movement to militarize the civilian aviation community of CAP into a fourth arm of the nation's defense unfolded over the first half of 1942. A small group of civilian volunteers, OCD's first members to actively engage enemy forces in defense of the homeland, became the fulcrum for the grand CAP experiment.

From the success of *Paukenschlag* in January, Doenitz ordered more waves of U-boats to American waters where they enjoyed continued success. Among industry and government officials, heavy tanker losses along the East Coast grew as a matter of grave concern. In the Eastern Sea Frontier, the limited number of ships and aircraft meant that Andrews had to disperse his assets rather than concentrate them at specific sectors. The former choice provided less-than-ideal defense and necessitated additional assistance to successfully fight the U-boats.[39] For I Bomber Command, almost 8,000 hours of patrol from January to February resulted in only four attacks with no evidence of damage or destruction of the enemy.[40] By the end of February, U-boat attacks had either damaged or destroyed 22 tankers.[41] The Navy assured the American people it had the situation under control, claiming to have sunk or damaged 21 submarines in the Atlantic since December 1941.[42] In reality, the Navy did not destroy a U-boat until one of its PBO-1 Hudson light patrol bombers hit and sank *U-656* off Newfoundland on 1 March; the Navy did not destroy a U-boat off the US East Coast until 14 April (*U-85*).[43] The uncoordinated and inadequate response by the Army and Navy left merchant traffic exposed and vulnerable, and Doenitz and his U-boats profited as a result. Within the first half of 1942, U-boats sank approximately 3 million tons of shipping in American waters at a cost of only eight submarines lost.[44]

One CAP pilot deeply aware of the terrible shipping losses in early 1942 was William D. Mason. In January, Mason, manager of Sun Oil's Marcus Hook Refinery in Pennsylvania, shared his views about CAP and civil aviation's potential with company president, J. Howard Pew. Mason believed CAP aircraft could fly patrols over the shipping lanes to protect tankers and shore up the tanker crews' uneasy morale. A U-boat spotting a CAP aircraft might hesitate to brazenly attack a tanker and risk aerial retaliation. Sun Oil already found itself struggling to keep crews on its tankers because the merchant seamen felt they were not adequately protected. Mason and Sun Oil believed CAP's coverage would boost crew morale immeasurably.[45] On 3 February, Mason shared his idea about coastal patrols in a conversation with Curry. The general said he required money and a directive to launch the effort; Pew guaranteed $10,000 in funds.[46]

Curry pursued the directive. He spoke with Lieutenant General Drum and Brigadier General Krogstad at I Bomber Command about using CAP aircraft on coastal patrol duty. Both gave Curry their informal concurrence. On 11 February, Curry met with CAP wing commanders of the Eastern Seaboard states in Washington, DC, to discuss coastal patrol arrangements. Saville addressed the group, bringing a message from Arnold that "the CAP would be used to the limit of its capabilities."[47] To provide coverage from Cape Henry,

Virginia, to Cape Hatteras, North Carolina, the wing commanders discussed what options for bases existed in conjunction with Army striking forces. For the striking force near Fort Dix, New Jersey, CAP designated bases at Atlantic City, New Jersey, and at Rehoboth, Delaware. For the southern area operating in conjunction with the striking force at Langley Field, Virginia, they recommended bases at Virginia Beach and at Manteo, North Carolina.[48] Wright "Ike" Vermilya Jr., Florida Wing Commander, did not attend, as Curry had already worked out a coastal patrol plan with him ready for execution at any time.[49]

A finalized CAP plan emerged for two patrol bases: The first base at Atlantic City would use six aircraft equipped with two-way radios flying daily pairs of two-hour missions patrolling 15 miles offshore from Barnegat Light to Cape May, New Jersey. A second base at Rehoboth featured two units of three aircraft each patrolling from Cape May, New Jersey, to Cape Henlopen, Delaware, and between Cape Henlopen and Chincoteague Inlet, Virginia. Each base required "a very responsible person" to run the facility and the operations. The base commander would be responsible for communications between the Army and managing pay vouchers, with base personnel kept to a minimum of at least six ground crew personnel. Obvious challenges to launching the effort included ferrying personnel from home to the bases. Since "there was nothing at all at Rehoboth Beach," the Delaware Wing commander had to provide a gasoline truck and see to the cleaning of the hangar.[50]

With the Air Staff and General Headquarters in the War Department favorably disposed to the CAP plan, the effort could commence immediately upon the authorization of Army Chief of Staff, Gen George C. Marshall. The actual details and overall general scheme of base operations remained to be drawn up by the respective wing commanders. Other CAP leaders in attendance inquired about operations in their states. Georgia wing commander Winship Nunnally of Atlanta asked about a patrol along the state's coastline, believing six aircraft at Savannah and three out of Brunswick would be enough. The meeting notes do not specify if Curry responded directly to Nunnally, but the Georgia wing commander was advised to ask his group commander in Savannah to "think about the possibility of operating coast patrols out of there" since "if the C.A.P. is properly organized and planned, I am sure the Army will give us definite clearance."[51]

After the meeting, Curry submitted a formalized, revised plan to Arnold. Without mentioning specific base locations, he proposed using

Figure 5. Letter from William D. Mason to Earle L. Johnson of 4 February 1942 offering Sun Oil funds for an experimental coastal patrol. (Document courtesy of the Barry L. Spink Papers via the Morse Center.)

his volunteer aviators for 30 days as an auxiliary force for the Army Air Forces antisubmarine patrol effort under control of the I Air Support Command. CAP units would be stationed at intervals along the coast from Sandy Hook, New Jersey, to Cape Hatteras. To provide this degree of coverage, Curry estimated the daily operational costs at $1,500 for 36 aircraft flying daytime patrols 15 miles out to sea. Communication plans would be worked out through the Support Command's existing setup. He mentioned Sun Oil's offer of $10,000 for the experimental mission and closed by requesting a sum of $45,000 from the Army to provide a 30-day initial experimental mission.[52]

Maj Gen Millard F. Harmon, chief of the Air Staff, forwarded Curry's plan to Army General Headquarters. Harmon explained that pilots would receive "some remuneration from the Government in order to place them on a military status"—Sun Oil's money would not be used. In his reply to Arnold about Curry's proposal, Lt Gen Leslie J. McNair, chief of staff, General Headquarters, US Army, wrote that he doubted "that the proposed employment of the Civil Air Patrol would be practicable and effective" but recommended the commanding general, Eastern Theater of Operations, be authorized to accept the services of the Civil Air Patrol on a small scale, for trial. The trial would be contingent on Curry presenting a plan of operations acceptable to Lieutenant General Drum.[53]

On 23 February, Lt Col Nathan F. Twining, secretary of the Air Staff, instructed the Army Air Forces Budget Section to designate funds for a 30-day CAP coastal patrol experiment to cover aircraft rental costs (fuel, oil, maintenance, parts, and depreciation) and per diem for aircrews. First Air Force would observe the operation and results of the patrol, and, "if found to be of sufficient value," additional CAP patrols "will probably be established."[54] To support Twining's endorsement of the experiment to the Army General Staff, Curry provided additional details on CAP's plan from the previous CAP discussions of 11 February. The New Jersey Wing would operate six aircraft from Atlantic City Air Base, patrolling from Barnegat Light to Cape May. Under the direction of the Dover Army Air Base, the Delaware Wing would operate seven aircraft from a grass field at Rehoboth from Cape May to Cape Henlopen. Both bases would operate under the direct control and command of the Army Air Base commander, flying such missions and on such schedules as he designated. CAP would observe all military rules and regulations in the area of operation.[55] Five days later, General Marshall authorized Curry to conduct the 30-day experiment.⁵⁶ CAP National Headquarters immediately activated its First and Second Task Forces, located at Atlantic City Municipal Airport, New Jersey; and Rehoboth Airport, Rehoboth, Delaware, respectively.⁵⁷

Additional support for using civilian aircraft in antisubmarine warfare soon found its way to Arnold's desk. In the wake of several publicized attacks by Japanese and German submarines off Santa Barbara, California, and Stuart, Florida, respectively, Assistant Secretary of War for Air Robert A. Lovett wrote Arnold and suggested calling for CAP aircraft for offshore patrols.⁵⁸ A pioneering member of First Yale Unit, he had served as a naval aviator in World War I and continued to fly light aircraft on occasion in the interwar years. An NAA member since 1928, Lovett believed in the potential military application of light civilian aircraft.⁵⁹ He postulated that the aircraft, "while not carrying any bombs or armament, would, nevertheless, if painted with Air Corps war paint and insignia, be apt to cause any enemy submarine to give birth to a set of tin dishes through fright. Their main purposes would, of course, be to notify the regular Air Forces of any suspicious vessels sighted." Using the aircraft at vulnerable locations, such as San Diego; San Francisco; Newport, Rhode Island; New York; Norfolk, Virginia; Miami; and ports in the Gulf of Mexico, although "obviously not an efficient answer to the problem . . . would at least give us the satisfaction of having used every bit of ingenuity possible in order to tide over the situation until specialized weapons are present in adequate quantities," concluded Lovett.⁶⁰ Arnold replied, being "heartily in accord with your suggestion that civilian planes be used for this purpose" and informed the assistant secretary of war that such planes were being used on the East but not West Coast.⁶¹

As CAP prepared its two task forces, word of the experiment spread among Washington, DC, bureaucracy. On 3 March, the Petroleum Industry War Council convened and discussed the tanker losses. The council appointed a Temporary Committee on the Protection of Tankers to meet with representatives of Army and Navy to discuss the feasibility of using CAP aircraft to better protect tankers from submarine attack. The following day, the temporary committee of five petroleum industry leaders (including Mason of Sun Oil) joined with six representatives from the Army and Navy to discuss tanker operation problems.

While the oilmen expressed their growing frustration at the military's tepid response to increasing tanker losses, the uniformed officers detailed Army and Navy progress to date. Members discussed the CAP and collectively endorsed an immediate and thorough test for coastal patrol work. By extrapolating losses based on the current loss rate of tankers from January to early March 1942, the nation stood to lose an estimated 125 tankers (of the current fleet of 320) and approximately 3,000 lives. This would leave the East Coast with only 10 million barrels of crude oil by year's end.[62] Such continued losses, if sustained, would be intolerable by 1943. The oilmen warned the military officials that the losses to date had demoralized crews and many, particularly engineers, were leaving.[63] Rough notes of the ensuing discussion included the possibility of establishing 18 air bases on the East Coast, 10 along the Gulf of Mexico, and 12 on the West Coast. The 40 bases would require a total of 800 personnel (pilots, observers, mechanics, and radio personnel) and 200 aircraft with two-way radios. Rough estimates for the cost of each base totaled $12,000 per day ($4,000 for personnel; $8,000 for aircraft), $360,000 per month, or $4,320,000 annually. A suggested alternative would be to provide no additional Army or Navy air protection for the tankers except the absolute minimum. Instead, the federal government could purchase or hire 200 civilian aircraft of 125 to 225 horsepower with two-way radios for coastal patrol duty, possibly equipped with 100-pound or heavier bombs and flotation gear for the aircrews.[64]

On 5 March, the first CAP coastal patrol flight took off from the grass field at Rehoboth. On that dark, hazy afternoon, two Fairchild Model 24 aircraft lifted off from their field and patrolled for an hour and 10 minutes. The aircraft flew roughly 20 miles offshore, then turned parallel to the coast, reporting nothing but water and a few oil slicks.[65] Three days later, I Air Support Command issued formal instructions for the two CAP task forces. The task forces were ordered "to establish an inshore anti-submarine patrol for the purpose of reporting the locations of enemy submarines and friendly vessels in distress." The commanding officer of the 59th Observation Group, with the cooperation of the 104th Observation Squadron, oversaw the experiment.[66]

The instructions provided by I Air Support Command would be repeated with subtle variation for all of CAP's coastal patrol bases. The First and Second Task Forces were instructed to provide maximum patrol coverage over their entire assigned sector, flying standing patrols from dawn to dusk no more than 15 miles offshore over the New Jersey and Delaware areas. Dawn patrols were to commence as soon as there was sufficient light for observation. Dusk patrols would remain airborne as late as possible, with additional patrols during daylight at irregular times. As a precaution, I Air Support Command instructed all aircraft not to approach closer than 1,500 yards from any surface vessel as recognition signals would not be made available to CAP aircrews and naval armed guards might accidentally mistake friend for foe. Aircrew were instructed to wear life vests while engaged in overwater operations and "every effort should be made to equip aircraft with a suitable life raft for additional safety."[67] All CAP aircraft would have functional two-way radios and remain in constant communication with a CAP-furnished ground station located at the airdrome. At intervals of not more than 30 minutes, position reports, code-named TR (Tare Roger), would be transmitted for each aircraft along with course, speed, and time by 24-hour clock for use in plotting the locations of the aircraft back at the base operations center. An example of a TR would read "TR 032 522 022 095 1415," or "aircraft at 40°32' North, 75°22' West, true course of 22° at ground speed 95 miles per hour at 2:15 p.m." In the event of radio failure, aircraft were ordered to immediately return to the nearest land and then back to base.[68]

Should the patrols observe anything, the aircrews would immediately radio in a contact report, a CR (Cast Roger). CRs would include the number and type of the observed item, latitude and longitude, approximate speed, time of observation, and estimated true course of the reported craft to the nearest 10 degrees. Destroyers were referred to as FG (Fox George); surfaced submarines, EF (Easy Fox); submerged submarines, ZU (Zed Units); tankers, BC (Baker Cast); aircraft, DE (Dog Easy); and unknowns, UT (Unit Tare). CAP coastal patrol aircrew received additional code phrases over the following weeks and months. A sampling of these code phrases included wreckage, PI (Prep Inter); survivors, SZ (Sail Zed); sinking, MO (Mike Option); aground, RQ (Roger Queen); stationary, LV (Love Victor); floating mine, AB (Affirm Baker); freighter, XH (X-ray Hypo); and enemy aircraft, EA (Easy Affirm). A potential CAP CR might read "WXBM from WXBM11 CAST ROGER 1 ZED UNIT 600 705 0746 3 90 WXBM11 Go Ahead." This would translate into "Second Task Force (Rehoboth) base from aircraft no. 11, contact report, one submerged submarine, latitude/longitude, time, estimated speed 3 knots, true course 90°, aircraft no. 11, end." The patrol contact reports would be radioed back to the CAP task force where a direct telephone line would connect the CAP bases with the Army's 104th Observation Squadron. Upon completion of the experiment, I Air Support Command would supply First Air Force a report on the desirability of using CAP in future submarine patrols.[69]

On the first patrol out of Atlantic City on 10 March, CAP demonstrated its utility. Approximately 15 minutes from base the patrol spotted the foundering hulk of the steam tanker *Gulftrade*, torn in two by a single torpedo fired earlier in the morning by *U-588*. The CAP aircrew reported observing the bodies of some of those killed in the attack as well as several survivors in the water. Over the course of the almost five-hour patrol the lone CAP aircraft located seven floating bodies, an empty lifeboat, and an apple crate.[70] Interviewed years later, Mason recalled how Atlantic City's first patrol "made up for the initial cost" of the CAP experiment.[71] That same day 20 miles east of Cape May at Five Fathom Bank shoal, two Fairchild 24s out of Rehoboth reported sighting a potential submarine preparing to attack two tankers. The aircraft dove towards the potential attacker, which disappeared, leaving the tankers unmolested to steam onward.[72] The closest U-boat at the time, *U-94*, made no mention of tankers but did report crash diving for "land-based aircraft, antiquated type, large and slow" in its *Kriegstagebüch* (KTB), or war diary.[73]

Getting the first two task forces operational provided an array of anecdotes illustrating the challenges CAP faced. To Army personnel in Atlantic City, the civilian aviators in their motley uniforms and varied, light aircraft seemed far from anything approaching military discipline. Wilson, overseeing the establishment of the First Task Force with its commander, Wynant C. Farr of Monroe, New York, recalled the result of one early patrol report. A Navy admiral unaware of CAP's new operation gave Wilson a tongue-lashing over the phone because of one CAP pilot's "scatterbrained report of a ship sinking and to cease civilian meddling in military affairs."[74]

At Rehoboth, Delaware Wing Commander Holger Hoiriis stood his task force up with the confidence. A native of Denmark, he came to the United States in 1923 intent on studying scientific farming methods but instead established a taxi service in New York and became a barnstorming pilot. In June 1931, he flew photographer Otto Hillig from Harbor Grace, Newfoundland, to Copenhagen, Denmark. For becoming the first aviator to fly a paying passenger across the

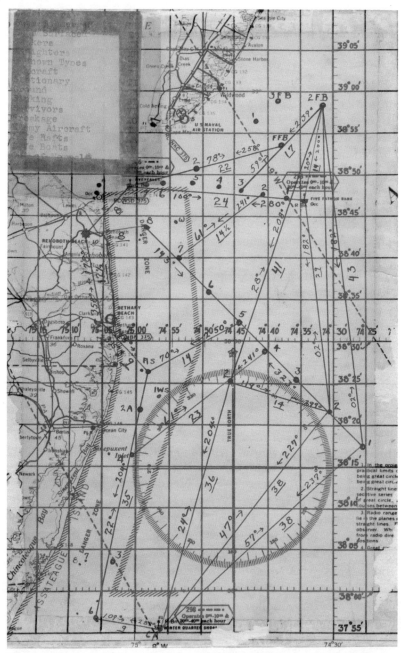

Figure 6. Sectional chart used for navigation by 1st Lt Henry E. Phipps while assigned to the Second Task Force, Rehoboth, Delaware, 1942–1943.
(Document courtesy of the Henry E. Phipps Papers via the Morse Center.)

Atlantic, King Christian X of Denmark knighted Hoiriis by conferring upon him the Order of the Dannebrog.[75] Although the aerial Danish knight initially encountered locals suspicious of civilians being able to fly when the CAA had grounded everyone else, Hoiriis swiftly assembled some of the best aircrews and ground personnel in the state. The ability to launch the first CAP coastal patrol flight owed much to his leadership and high state of organization of the wing.[76]

Figure 7. Members of the First Task Force, Atlantic City, New Jersey, 13 March 1942, sporting an array of life vests and uniform variations. Gill Robb Wilson is ninth from the left. (Photograph courtesy of the Morse Center.)

Once made aware of the Army's CAP experiment, the Navy began to take a closer look at the civilian operation. On 12 March, Andrews wrote to King, now dual-hatted as chief of naval operations and COMINCH, about using a "scarecrow patrol" of aircraft in a continuous daylight patrol off the entire coast, akin to Lovett's thoughts for CAP. Andrews explained that "the more planes that can be used in patrolling during daylight hours, the greater will be the chances of keeping down enemy submarines. It has been found that these submarines upon sighting any plane, dive and submerge immediately." Andrews mentioned how he "had in mind for some time" using CAP aircraft for continuous daylight patrols for limited distances off the entire coast. These planes would form his scarecrow patrols and would disband once the Navy had sufficient combat planes for the same purpose. Andrew requested King's approval for this proposal and full authority to immediately make such arrangements.⁷⁷

Figure 8. Flight line at the First Task Force, Atlantic City, New Jersey, showcasing a variety of civilian airframes. (Photograph courtesy of Charles B. Compton via the Morse Center.)

Deputy Chief of Staff for Operations Rear Adm Richard S. Edwards provided King with insight. As a participant in the Tanker Protection Committee meeting of 4 March, Edwards informed King that Maj Gen Carl Spaatz, chief of the Army Air Force Combat Command, mentioned the War Department's acceptance of CAP for scarecrow patrol use. Writing King, Edwards expressed his opinion that the CAP flights "will serve no useful purpose except to give merchant ships the illusion that an adequate air patrol is being maintained." He also shared with King the opinion of Rear Adm Donald B. Duncan, assistant chief of staff, that CAP's scheme was promoted by the builder of pleasure aircraft.⁷⁸ Edwards added how Duncan also felt CAP would detrimentally clog communications with false contact reports and mentioned "the probability that lost amateur flyers will require the use of anti-sub vessels to look for them."⁷⁹ Taking Edwards's input into account, King replied to Andrews to reject the use of civilian aircraft for coastal patrol duty. In his words, CAP's aircraft "would not be productive in sufficient degree to compensate for the operational difficulties to be encountered in coordinating and controlling the flying involved by inexperienced personnel." He closed by assuring Andrews of actions underway to improve sea frontier naval air activities as soon as aircraft and personal became available.[80]

Despite King's dismissal of CAP's potential, the Army Air Forces wanted more. In mid-March, Arnold asked Connolly to "take the necessary steps at once to organize the civilian puddle jumper pilots into squadrons for Coast Defense patrol work."[81] A week later, First Air Force commander Maj Gen Follett Bradley recommended establishing a CAP coastal patrol unit on the Delmarva (Delaware-Maryland-Virginia) peninsula to allow the relocation of Army units to improve the air defense of the Cape Hatteras area.[82] Curry, mindful of the stakes at hand, recognized an expansion required even more discipline. The day before the first coastal patrol mission, he issued an operations directive reminding every member that "we are at war" and "without thorough air discipline, the Civil Air Patrol is of no value as a flying auxiliary to the armed forces."[83]

At Atlantic City, Maj Charles A. Masson, commanding the 104th Observation Squadron, found that CAP's personnel provided a great value and close coverage of its assigned patrol area. The first week of CAP patrols had as many contact reports as the squadron had amassed in two months of operation. CAP's close coverage along the New Jersey coast enabled Masson to concentrate Army assets in other areas of concern. He believed that, rather than end as an experiment, the coastal patrols should become permanent.[84] On 31 March Twining authorized the expenditure of funds previously designated on 23 February for the 30-day experiment, but also "for an additional period of 30 days or so long as any of such funds remain unexpended."[85] The Army Air Forces transferred $40,000 to the Treasury Department for disbursement by CAP for per diem allowances, aircraft expenditures, and other incidental expenses.[86]

By March 1942, the need for better coordination between Army and Navy antisubmarine operations pulled CAP's effort into the Navy orbit. On 23 March, Secretary of the Interior Harold L. Ickes expressed concerns to Roosevelt about the continued loss of tankers and the need for a unified patrol effort between the armed forces. The president in turn asked King and Marshall if anything was being done on the matter.[87]

Both service chiefs agreed three days later that the sea frontier commanders would exercise jurisdiction for all naval forces and Army aviation units engaged in antisubmarine operations and protection of shipping.[88] On 28 March, Drum wrote to Andrews and allocated to him elements of the I Bomber Command, I Air Support Command, and all CAP units operating under I Air Support Command.[89] Despite King's dismissal of CAP, Andrews was now positioned to implement his scarecrow patrol along the nation's shorelines.

Concurrently in late March–early April, representatives from the Bureau of the Budget interviewed Andrews and Army representatives for their personal views on the CAP coastal patrol effort. Andrews encouraged the use of CAP by the Army and believed the civilian effort provided valuable intelligence. Even unarmed, he asserted, CAP served effectively as a scarecrow patrol, and use of the CAP aircraft should be continued until military forces could take over the job. Andrews saw no need to change CAP from a civilian to a military organization. Notably, Andrews told the bureau representatives that "the C.A.P. has so far exercised fully adequate control and discipline of its members." The admiral reported writing a letter recommending the Navy Department release an official policy statement in favor of utilizing CAP services, although he had yet to receive a reply.[90]

Two officers from I Air Support Command, Lt Col John P. Doyle and Lt Col John E. Bodle, spoke at length about the coastal patrol effort. Both felt a great need existed for using CAP due to military necessity and that CAP's deterrent usage was its greatest value, concluding that "the greater the number of eyes we have in the air, the fewer sinkings we are going to have." Noting that CAP "had not failed in a single instance to fully accomplish all missions assigned" and experienced no "dereliction or deficiency in the performance of duty," the men supported continuing CAP's civilian effort. Like Andrews, these officers felt CAP should remain a civilian effort. Because CAP reemployed pilots who "can not possibly meet military standards and therefore can not be recruited in the Army or Navy as pilots," CAP units freed up military personnel for service elsewhere.[91]

The Budget Bureau concluded that CAP's coastal mission was useful and performed in a satisfactory manner. Unfortunately, no conclusion could be reached to determine how long CAP would be needed for coastal patrol service, at least not until sufficient military aircraft were available. The Bureau decided then that no action be taken to disrupt the operation without the concurrence of the War and Navy Departments. Regarding funding, the bureau wrote that consideration would "be given to a system of allocation of funds for specific C.A.P. missions by the Department which calls upon C.A.P. for service." The funding procedure would in turn "give a degree of control" over CAP that would not be possible if CAP received allocations directly.[92] The War and Navy Departments would need to reach definite policy decisions for the contemplated use of CAP to provide the basis for program and financial planning.[93]

With the bases at Atlantic City and Rehoboth providing viable results, CAP National Headquarters activated a Third Task Force. On 29 March, Headquarters I Air Support Command issued instructions for the CAP squadron at Morrison Field, West Palm Beach, Florida, to conduct offshore patrols from Lake Worth Inlet to Melbourne Beach. CAP National Headquarters tapped Maj Ike Vermilya as the base commander; he promptly activated the Third Task Force the following day. The unit comprising the patrol was practically tailor made for the task. Almost a year before on 28 May 1941, Brig Gen Vivian Collins, Florida's Adjutant General, had mustered the 1st Air Squadron of the Florida Defense Force under Vermilya's capable leadership. A World War I veteran and former member of the Arkansas National Guard, Vermilya had over 12,000 flying hours to his credit and was a respected member of the Florida aviation community, and he assembled a capable fleet of aircraft and experienced pilots. The volunteer unit provided an aerial element to the state's civilian defense effort. Members received training like the men of the Florida National Guard, including infantry drill and the handling of small arms. Named as CAP's Florida Wing Commander on 10 December, Vermilya already had a fully operational, military-trained and -uniformed squadron in existence. The staff of the 1st Air Squadron shortly became Vermilya's headquarters staff, and by February 1942 the wing featured seven groups and 22 squadrons performing formation flights, simulated missing aircraft patrols, and coastal patrol flights for the state defense force.[94]

Gill Robb Wilson flew down from Atlantic City to conduct a three-day instructional course for the pilots and observers based on the previous weeks' experiences. Drawing extensively from his seasoned 1st Air Squadron personnel, Vermilya began active patrols on 2 April. On its first day, the Third Task Force launched 15 planes conducting

14 missions for approximately 42 hours of reconnaissance patrols. Eleven days after commencing patrols, aircraft from the base assisted the Coast Guard in locating survivors and bodies from the Swedish freighter *Korsholm*, which had been shelled and sunk by *U-123* off Cocoa Beach.[95] This performance by the Third Task Force notably contradicted King's concern about "operational difficulties."

CAP's three bases, without expending military resources beyond personnel coordination, had thus far freed Army Air Forces assets to conduct missions in other operational areas. The civilian aircrews and base personnel demonstrated military discipline and competency on a small scale, drawn from wings with strong organizations. What remained unknown was CAP's ability to expand and sustain coastal patrol operations for extended periods of time.

Notes

Epigraph. Doenitz, *Memoirs*, 202.

---

1. War Diary, North Atlantic Naval Coastal Frontier, December 1941, chap. 2, 10–11, NARA (via Fold3); and Gannon, *Drumbeat*, 176–77. Under the orders of Adm Harold R. Stark, Chief of Naval Operations, the Navy established Coastal Frontiers on 1 July 1941. These were areas of command that began at the shoreline of the United States and extended outward approximately 200 miles. The Coastal Frontiers were larger operational areas than the individual Naval Districts and included not just the Districts but the shipping lanes within the sea area. A Coastal Frontier commander had operational control of all allocated forces within his jurisdiction. On 1 February 1942, the Coastal Frontiers were renamed Sea Frontiers. At the onset of the war, the North Atlantic Naval Coastal Frontier (later ESF) encompassed the area from the North International Boundary (Maine-Canadian border) to the lower line of Onslow County, North Carolina. In early February 1942, the ESF expanded southward to the boundary between Duval and St. Johns counties, Florida. Morison, *Battle of the Atlantic*, 71n20, 207–8; and Gannon, *Drumbeat*, 171.

2. War Diary, Eastern Sea Frontier, February 1942, chap. 3, 1, NARA (via Fold3).

3. First Air Force originally was activated as the Northeast Air District of the General Headquarters Air Force on 18 December 1940 as one of the original four numbered air forces. It was redesignated as First Air Force on 9 April 1941, tasked with the defense of the Northeast and the Great Lakes regions of the United States. At the onset of World War II, First Air Force bore responsibility for air defense of the entire Eastern Seaboard. Continental United States NORAD (North American Aerospace Defense Command) Region–First Air Force (Air Forces Northern), "1st AF Mission," 25 February 2013, https://www.1af.acc.af.mil/Library/Fact-Sheets/Display/Article/289618/1st-af-mission/; and Air Force Historical Research Agency, "First Air Force (Air Forces Northern) (ACC)," 11 June 2009, https://www.afhra.af.mil/About-Us/Fact-Sheets/Display/Article/433194/first-air-force-air-forces-northern-acc/. On 24 June 2016, a realignment resulted in the transfer of CAP-USAF from Air Education and Training Command to Air Combat Command, with First Air

Force once again approving all CAP operational missions. Air Combat Command, "1st Air Force Participates in Civil Air Patrol Transfer of Authority Ceremony," 24 June 2016, https://www.acc.af.mil/News/Article-Display/Article/815203/1st-air-force-participates-in-civil-air-patrol-transfer-of-authority-ceremony/.

4. Gannon, *Operation Drumbeat*, 182–84; Craven and Cate, *Plans and Early Operations*, 521–27; and Blair, *Hitler's U-Boat War: The Hunters, 1939-1942*, 461-64. All subsequent references to this volume appear as Blair, *U-Boat War*.

5. Blair, *U-Boat War*, 481.

6. War Diary, Eastern Sea Frontier, February 1942, chap. 3, 2–4, NARA (via Fold3); War Diary, Eastern Sea Frontier, March 1942, chap. 6, 1–2, NARA, (via Fold3); Blair, *U-Boat War*, 498; Roskill, *The Period of Balance*, vol. 2, *The War at Sea*, 97; and Hickam, *Torpedo Junction*, 199–208.

7. Craven and Cate, *Plans and Early Operations*, 524.

8. Gannon, *Drumbeat*, 385–86; Blair, *U-Boat War*, 447–52, 455–60; Roskill, *The Period of Balance*, vol. 2, *The War at Sea*, 97–99; Baer, *One Hundred Years of Sea Power*, 194–98; and Hickam, *Torpedo Junction*, 116–18.

9. Frank A. Flynn to John F. Curry, 6 January 1942, BLS, CAP-NAHC.

10. Henry King to Reed G. Landis, 30 January 1942, BLS, CAP-NAHC.

11. Anthony Schmittinger to John F. Curry, undated (February 1942); Robert Taylor to Anthony Schmittinger, 16 February 1942, Reel 38919, AFHRA.

12. Irving H. Taylor to John F. Curry, 24 January 1942, BLS, CAP-NAHC.

13. John F. Curry to Henry H. Arnold, memorandum, subject: Brief Report on Activities of Civil Air Patrol, 23 January 1942, binder "Legal Status, Administrative Concepts, and Relationship of the Civil Air Patrol, 1941 to 1949," CAP-NAHC. By 30 January 1942, CAP membership applications totaled 20.8 percent of all registered pilots nationwide, with overall membership at approximately 17,000. CAPNHQ, "Percentage of Applications Received by States," 30 January 1942; and Reed G. Landis to John F. Curry, memorandum, 29 January 1942, BLS, CAP-NAHC.

14. John F. Curry to Henry H. Arnold, memorandum, subject: Employment of Civil Air Patrol on Pacific Coast, 31 January 1942, BLS, CAP-NAHC. On 31 January, the Eastern Sea Frontier adjusted coastal shipping routes to bring shipping as close to the shoreline as safe navigation permitted, with instructions for ships to sail at night along said routes with their navigation lights turned off. War Diary, Eastern Sea Frontier, March 1942, chap. 2, 1, NARA (via Fold3).

15. CAPNHQ, Groups and Squadrons Organized to Date: Missions Equipped, 5 February 1942, BLS, CAP-NAHC.

16. Henry H. Arnold to Donald H. Connolly, memorandum, subject: Employment of Planes and Pilots, 4 February 1942, Folder "Military Official Nos. 49–54," Box 167, HHA, LOC.

17. John F. Curry to Hugh A. Drum, 7 February 1942, BLS, CAP-NAHC.

18. Hugh A. Drum to John F. Curry, 13 February 1942, BLS, CAP-NAHC.

19. Ritchie, *James M. Landis: Dean of the Regulators*, 106–8; Dallek, *Defenseless Under the Night*, 206–7; and James M. Landis to Wayne Coy, 28 January 1942, Folder "OCD Civil Air Patrol," Box 10, Entry 107A, RG51, NARA.

20. John F. Curry to James M. Landis, memorandum, subject: Funds for Civil Air Patrol, 25 January 1942, BLS, CAP-NAHC.

21. James M. Landis to Wayne Coy, 28 January 1942, Folder "OCD Civil Air Patrol," Box 10, Entry 107A, RG51, NARA.

22. John F. Curry to Henry H. Arnold, memorandum, subject: Civil Air Patrol Budget (with attachment), 27 January 1942, BLS, CAP-NAHC.

23. Henry H. Arnold to John F. Curry, memorandum, subject: Civil Air Patrol Budget, 4 February 1942, BLS, CAP-NAHC.

24. Ivan Hinderaker to Bernard L. Gladieux, memorandum, subject: Conference Report on the Civil Air Patrol Request for $17,142 for registering qualifications of civilian airplane pilots, and for $97,000 to provide administrative assistance to State Civil Air Patrol Wing Commanders, 10 February 1942, Folder "OCD Civil Air Patrol," Box 10, Entry 107A, RG51, NARA.

25. Herman Kehrli to Bernard L. Gladieux, memorandum, subject: Civil Air Patrol, 6 March 1942, Folder "OCD Civil Air Patrol," Box 10, Entry 107A, RG51, NARA; and John F. Curry to Bertrand Rhine, 12 February 1942, BLS, CAP-NAHC.

26. Reed G. Landis to Gordon P. Saville, 11 February 1942, with attachment, BLS, CAP-NAHC.

27. Ivan Hinderaker to Bernard L. Gladieux, memorandum, subject: Conference with Lt. Colonel Saville, Officer Designated to Handle Army Relationships with the Civil Air Patrol, 18 February 1942, Folder "OCD Civil Air Patrol," Box 10, Entry 107A, RG51, NARA.

28. Ivan Hinderaker to Bernard L. Gladieux, memorandum, subject: Conference with Lt. Colonel Saville, Officer Designated to Handle Army Relationships with the Civil Air Patrol, 18 February 1942; and Ivan Hinderaker to Bernard L. Gladieux, memorandum, subject: Army Attitude Toward the Civil Air Patrol, 18 February 1942, Folder "OCD Civil Air Patrol," Box 10, Entry 107A, RG51, NARA.

29. Link, *Army Air Forces Historical Studies No. 19*, 4.

30. Link, 59–67.

31. Dallek, *Defenseless Under the Night*, 175.

32. John F. Curry to Bertrand Rhine, 12 February 1942, BLS, CAP-NAHC.

33. OCD, CAPNHQ, Operations Directive No. 6, Table of Organization for Civil Air Patrol, 10 March 1942; and document, "Tentative Table of Organization–Civil Air Patrol," 25 February 1942, Reel 38907, AFHRA.

34. Landis received orders in March to report to active duty with the Army Air Forces, serving for the war in the Office of the Chief of Air Forces and with the Troop Carrier Command. Hudson, "Reed G. Landis," 140; and OCD, *CAP Bulletin*, 1, no. 8, 10 March 1942, Folder 6, Box 6, ELJ, WRHS.

35. Reed Landis to James M. Landis, memorandum, subject: Future of Civil Air Patrol, 9 March 1942, BLS, CAP-NAHC.

36. "Specialist Reserve" refers to the US Army Specialist Corps.

37. Reed G. Landis to John F. Curry, memorandum, 13 March 1942, BLS, CAPNAHC.

38. Reed G. Landis to John F. Curry, memorandum, 13 March 1942, BLS, CAPNAHC.

39. War Diary, Eastern Sea Frontier, February 1942, chap. 3, 8, NARA (via Fold3).

40. Craven and Cate, *Plans and Early Operations*, 527.

41. Offley, *Burning Shore*, 118; Frey and Ide, *A History of the Petroleum Administration for War*, 87; and Morison, *Battle of the Atlantic*, 130–33.

42. Offley, *Burning Shore*, 122–23; "U.S. Navy Sinks or Damages 21 Hitler U-boats: 70 Subs Attacked, Knox Reveals in 'Conservative Report,'" *The Pittsburgh (PA) Press*, 26 February 1942, 2; John G. Norris, "119 Jap Vessels Is U.S. Toll; 3 Enemy Subs Sunk in Atlantic," *Washington Post*, 26 February 1942, 1; and Dimbley, *Battle of the Atlantic*, 250–51.

43. Hickam, *Torpedo Junction*, 137–57; and Blair, *U-Boat War*, 541–44.

44. Murray and Millett, *A War to Be Won*, 251; and Cressman, *Official Chronology of the U.S. Navy in World War II*, 80, 88.

45. John F. Curry to Henry H. Arnold, draft memorandum, subject: Use of the Civil Air Patrol as an Auxiliary to the Air Force Coastal Patrol, undated (believed 11

February 1942), Reel 38920, AFHRA. Over the course of the war, Sun Oil lost four tankers to enemy attacks (SS *J. N. Pew*, MS *Sunoil*, MS *Mercury Sun*, and MS *Atlantic Sun*) while three others suffered damage from torpedo strikes. A total of 141 merchant seamen and 36 Naval Armed Guardsmen died aboard Sun Oil tankers in the Battle of the Atlantic. Booklet, Sun Oil Company, "Marine Department," http://www.fleetsheet.com/sunmarine.pdf; and "Sun Oil to Erect Monument to 141 Seamen Lost in War," *Delaware County Daily Times* (Chester, PA), 17 February 1949, 1.

46. William D. Mason to John F. Curry, 4 February 1942; Earle L. Johnson to William D. Mason, 7 February 1942, BLS, CAP-NAHC; and House Committee on Education and Labor, *United States Compensation Act Benefits for Members of the Civil Air Patrol: Hearings on H.R. 3673*, 80th Cong., 2nd sess., 1948, 18.

47. John F. Curry to Bertrand Rhine, 12 February 1942, BLS, CAP-NAHC.

48. Minutes of the Meeting of the Wing Commanders in Capt. [Robert] Taylor's Office on Wednesday, 11 February 1942, Reel 38918, AFHRA.

49. John F. Curry to Bertrand Rhine, 12 February 1942, BLS, CAP-NAHC.

50. Minutes of the Meeting of the Wing Commanders in Capt. [Robert] Taylor's Office on Wednesday, 11 February 1942, Reel 38918, AFHRA.

51. Minutes of the Meeting of the Wing Commanders in Capt. [Robert] Taylor's Office on Wednesday, 11 February 1942, Reel 38918, AFHRA.

52. John F. Curry to Henry H. Arnold, draft memorandum, subject: Use of the Civil Air Patrol as an Auxiliary to the Air Force Coastal Patrol, undated (believed 11 February 1942), Reel 38920, AFHRA; and Army Air Forces Antisubmarine Command, 7 December 1941 to 15 October 1942, Vol. II, Appendix, Section 20, "I Bomber Command," 7, Reel A4048, AFRHA.

53. Millard F. Harmon to C. W. Russell, memorandum, subject: Employment of Civil Air Patrol for Coast Patrol, 11 February 1942; and Leslie J. McNair to Henry H. Arnold, memorandum, subject: Employment of Civil Air Patrol for Coast Patrol, 17 February 1942, Folder "SAS 373," Box 110, HHA, LOC.

54. Nathan F. Twining to Budget Section, memorandum, subject: Coast Patrol by Civil Air Patrol, 23 February 1942, Folder "Military Miscellaneous #5," Box 206, HHA, LOC. The commanders of the first three task forces were authorized to spend up to $12,000 each for personnel and aircraft. Earle L. Johnson to Edward W. Bell, 15 April 1942, BLS, CAP-NAHC.

55. John F. Curry to Nathan F. Twining, memorandum, 23 February 1942, Reel 38918, AFHRA.

56. R. C. Lewis to Commanding General, Field Forces, memorandum, subject: Employment of Civil Air Patrol for Coast Patrol–1st Indorsement, 28 February 1942; J. W. Ramsey to Commanding General, Eastern Theater of Operations and First Army, memorandum, subject: 2nd Indorsement, 28 February 1942, Folder "SAS 373," Box 110, HHA, LOC; and C. H. Danielson to Commanding General, Air Forces, Eastern Theater of Operations and First Air Force, memorandum, subject: 3rd Indorsement, 6 March 1942, Reel A4064, AFHRA.

57. OCD, Civil Air Patrol, Operations Orders No. 1, Activation of CAP Coastal Patrols, 30 November 1942, Folder 2, Box 6, ELJ, WRHS.

58. The attacks Lovett referenced are most likely the shelling of the Ellwood Oil Field by the Imperial Japanese Navy submarine *I-17*, and the torpedoing of the steam tanker *W.D. Anderson* on 23 February 1942 by German submarine *U-504*.

59. Jordan, *Robert A. Lovett*, 8–25. One of Lovett's fellow members of the First Yale Unit was John Vorys.

60. Robert A. Lovett to Henry H. Arnold, memorandum, 25 February 1942, Folder "Military Official Nos. 107–110," Box 169, HHA, LOC.

61. Henry H. Arnold to Robert A. Lovett, memorandum, 26 February 1942, Folder "Military Official Nos. 270–273," Box 177, HHA, LOC.

62. Minutes of meeting of the Temporary Committee on Protection of Tankers, Petroleum Industry War Council with attached memorandum on East Coast transportation, 4 March 1942, Folder "Protection of Tankers, War Council (Temporary)," Box 690, Entry 16, Petroleum Industry War Council, Subcommittee Minutes–Minutes of the PIWC National and Special Committees, Dec 1941–Jan 1946, Record Group 252, Records of the Petroleum Administration for War, Petroleum Industry War Council, Subcommittee Minutes, Minutes of PIWC National and Special Committees, Dec 1941–Jan 1946, NARA. The petroleum industry representatives included B. B. Howard, chairman, Charles H. Kunz, William D. Mason, H. G. Schaad, and A. E. Watts. The military representatives included Undersecretary of War Robert P. Patterson, Undersecretary of the Navy Ralph Bard, Assistant Secretary of Air for War Robert A. Lovett, Maj Gen Carl Spaatz, Brig Gen Walter B. Pyron, and Rear Adm Richard S. Edwards. The document incorrectly lists Spaatz as a brigadier general and Edwards as a full admiral.

63. War Diary, Eastern Sea Frontier, April 1942, chap. 7, 3–4, NARA (via Fold3).

64. Document marked "Notes Taken for Capt. Taylor on 4 March 1942," Reel 38918, AFHRA.

65. Hugh R. Sharp Jr. to Robert Neprud, 23 January 1947; Hugh R. Sharp Jr. to Robert Neprud, 28 November 1946, Reel 38910, AFHRA; Edwards, interview; and Sharp, interview.

66. Headquarters I Air Support Command to Civil Air Patrols Authorities, Letter of Instructions Number 1, 8 March 1942, with attachments, Reel A4064, AFHRA.

67. Headquarters I Air Support Command to Civil Air Patrols Authorities, Letter of Instructions Number 1, 8 March 1942, with attachments, Reel A4064, AFHRA.

68. Headquarters I Air Support Command to Civil Air Patrols Authorities, Letter of Instructions Number 1, 8 March 1942, with attachments, Reel A4064, AFHRA.

69. Headquarters I Air Support Command to Civil Air Patrols Authorities, Letter of Instructions Number 1, 8 March 1942, with attachments; Headquarters I Air Support Command to Civil Air Patrol Authorities, Changes No. 1 to Letter of Instruction No. 1, 15 March 1942; Headquarters I Air Support Command to Civil Air Patrol Authorities, Letter of Instructions Number 3, 9 May 1942, Reel 38920; and H. B. Sepulveda to Commanding General, I Air Support Command, memorandum, subject: 4th Indorsement, 7 March 1942, Reel A4064, AFHRA.

70. Neprud, *Flying Minute Men*, 12; Mellor, *Sank Same*, 54–60; and notes of interview of Wynant C. Farr by Robert Neprud, circa 1946–47, Reel 38910, AFHRA. While Neprud in *Flying Minute Men* claims Farr and Muthig spotted survivors, news reports of the sinking make no mention of aircraft sighting survivors. "19 Still Missing on Ship Cut in Half Off Barnegat," *Asbury Park Press* (NJ), 11 March 1942, 1–2; and Alexander Kendrick, "Phila. Ship Torpedoed in View of Barnegat; 16 Men Saved, 19 Lost," *Philadelphia Inquirer*, 11 March 1942, 1.

71. Notes of interview of William D. Mason by Robert Neprud, circa 1946–47, Reel 38910, AFHRA.

72. Neprud, *Flying Minute Men*, 12; Hugh R. Sharp Jr. to Robert Neprud, 28 November 1946; and Rehoboth Beach fact sheet created by Robert Neprud, circa 1946–1947, Reel 38910, AFHRA.

73. A plot of the boat's daily position report places the submarine roughly 50 nautical miles away from the area of the CAP report. KTB for Eighth War Patrol of *U-94*, entry for 10 March 1942, http://uboatarchive.net/U-94/KTB94-8.htm.

74. Wilson, *I Walked with Giants*, 285–87; and Mellor, *Sank Same*, 41–46.

75. Frebert, *Delaware Aviation History*, 73; "Holger Hoiriis, Aviator, Dies; Was Ill a Year," *News Journal* (Wilmington, DE), 8 August 1942, 1, 2; "Tremendous Ovation for Danish Flyers," *Boston Globe*, 27 June 1931, 3; and Mellor, *Sank Same*, 67.

76. Sharp, interview; and Keefer, *From Maine to Mexico*, 37–38, 50.

77. Adolphus Andrews to Ernest J. King, memorandum, subject: Scarecrow Patrol, 12 March 1942, file "A4-2," Box 245, Headquarters, COMINCH, 1942–Secret, A4-1 to A4-3, Record Group 38, Records of the Office of the Chief of Naval Operations (RG38), NARA.

78. Adolphus Andrews to Ernest J. King, memorandum, subject: Scarecrow Patrol, 12 March 1942, file "A4-2," Box 245, Headquarters, COMINCH, 1942–Secret, A4-1 to A4-3, Record Group 38, Records of the Office of the Chief of Naval Operations (RG38), NARA.

79. Richard S. Edwards to Ernest J. King, memorandum, 15 March 1942, file "A4-2," Box 245, Headquarters, COMINCH, 1942–Secret, A4-1 to A4-3, RG38, NARA.

80. Ernest J. King to Adolphus Andrews, memorandum, subject: Scarecrow Patrol, 17 March 1942, file "A4-2," Box 245, Headquarters, COMINCH, 1942–Secret, A4-1 to A4-3, RG38, NARA.

81. Henry H. Arnold to Donald H. Connolly, memorandum, subject: Organization of Civilian Puddle Jumper Pilots, 16 March 1942, Folder "Military Official Nos. 154–160," Box 170, HHA, LOC.

82. Follett Bradley to Henry H. Arnold, memorandum, subject: Anti-submarine Defense in Vicinity of Cape Hatteras, 24 March 1942, Folder "SAS 370.31," Box 107, HHA, LOC. Bradley's recommendation for another CAP coastal patrol base most likely resulted in the authorization and establishment of a fourth base at Parksley, Virginia.

83. OCD, CAPNHQ, Operations Directive No. 4, Air Discipline, 4 March 1942, Reel 38907, AFHRA.

84. Robert Taylor III to St. Claire Streett, 16 March 1942, Reel 38918, AFHRA.

85. Nathan F. Twining to Budget Section, memorandum, subject: Coast Patrol by Civil Air Patrol, 31 March 1942, Folder "Military Miscellaneous #5," Box 206, HHA, LOC.

86. Nathan F. Twining to James M. Landis, 2 April 1942, Binder "CAP Historical Research-Policy File NR 1 1941-1945," CAP Historical Foundation Collection (CHF), CAP-NAHC.

87. Harold L. Ickes to Franklin D. Roosevelt, 23 March 1942; and Franklin D. Roosevelt to Ernest J. King and George C. Marshall, memorandum, 24 March 1942, Folder "Military Official No. 270–273," HHA, LOC.

88. Craven and Cate, *Plans and Early Operations*, 528.

89. Hugh A. Drum to Adolphus Andrews, memorandum, subject: Allocation of Army Aircraft for Operations Against Enemy Seaboard Activities, 28 March 1941, Folder "SAS 320.2," Box 83, HHA, LOC. Drum retained control of the 20th Bomb Squadron (Heavy), 4th Bomb Squadron (Medium), 80th Bomb Squadron (Medium), 9th Observation Squadron (Medium), 14th, 18th, and 19th Observation Squadron (Light).

90. E. R. Baker to L. C. Martin, memorandum, subject: Employment of the Civil Air Patrol by the Army and the Navy in the Inshore Coastal Patrol on the Atlantic Coast, 8 April 1942, Folder "OCD Civil Air Patrol," Box 10, Entry 107A, RG51, NARA.

91. Baker to Martin, memorandum.

92. Baker to Martin, memorandum.

93. Ivan H. Hinderaker to Bernard L. Gladieux, memorandum, subject: Utilization of CAP by Army and Navy on Inshore Atlantic Coastal Patrol, 10 April 1942, Folder "OCD Civil Air Patrol," Box 10, Entry 107A, RG51, NARA.

94. Reilly, "Florida's Flying Minute Men," 417–24; Keil, "Coastal Patrol No. Three, Lantana, Florida"; and CAPNHQ, Groups and Squadrons Organized to Date: Missions Performed, 5 February 1942, BLS, CAP-NAHC.

95. Headquarters I Air Support Command to Civil Air Patrols Authorities, Letter of Instructions Number 2, 29 March 1942, with attachments, Reel A4064, AFHRA; OCD, Civil Air Patrol, Operations Orders No. 1, Activation of CAP Coastal Patrols, 30 November 1942, Folder 2, ELJ, WRHS; Reilly, "Florida's Flying Minute Men," 426; Keil, "Coastal Patrol No. Three"; and Gannon, *Operation Drumbeat*, 376–77. Postwar, Lantana Field became Palm Beach County Airport. Due to overcrowding, the Third Task Force moved on 19 May 1942 from Morrison Field to Lantana Field. Orders for the base commander, Wright Vermilya, did not actually arrive until 11 April. Cecil Johnson to Wright Vermilya, 11 April 1942, Reel 38918, AFHRA.

Chapter 4

Learning and Expansion

Although we, ourselves, did not know too much, we at least knew that if we wrote down what the men were to do they would be bound to understand much quicker and be able to do it much better. At Atlantic City, we had learned by the trial and error method. At Parksley, we benefited by those trials and errors.

—Maj Isaac Burnham II, Fourth Task Force, Parksley, Virginia

From the first missions of March 1942, CAP coastal patrols proved useful to military officials. Flying slow, low-level patrols over the ocean, the unsophisticated CAP aircraft were ideal for spotting small objects easily missed by faster military aircraft.[1] CAP's aircraft provided a cheap and conveniently visible deterrent to U-boat surface operations. The low speed and small size of the civilian aircraft posed the greatest threat to U-boats by being difficult to sight by the watch crews. With submarine pressure hulls vulnerable to damage from bombs, U-boat men could ill afford any aerial surprise.

U-boat doctrine directed two options upon spotting an aircraft during the day. If a watch crew sighted an aircraft far in the distance, then the boat would change course, reduce speed, and turn away, showing a narrow outline while minimizing its visible wake. If an aircraft was flying directly toward a U-boat, the boat had to crash dive at once. This entailed submerging as quickly as possible and fleeing the area in case of retaliation, thereby breaking off potential attacks.[2] As Rear Admiral Andrews explained to representatives from the Bureau of the Budget, forcing a U-boat to dive and hide would enable potential targets to safely escape attack.[3] Aerial interruptions cost U-boats precious fuel—diesel for the boat and caloric for the crew—and robbed the enemy of the element of surprise.

One man at CAP National Headquarters oversaw the civilian coastal patrol effort: Col Harry H. Blee. The California native received a commission as a captain in the Army Air Service in World War I and served in the Airplane Engineering Division at McCook Field in Dayton, Ohio. From 1927 to 1933, he served as the director of aeronautic development for the Department of Commerce, overseeing the department's research and development work on aircraft, engines, airports, air traffic control, air navigational charts, visual and radio aids for air navigation, radio-directed landing systems, and public relations work. Throughout the Roosevelt administration, Blee worked in private industry as a consulting aeronautical engineer. Promoted to colonel in the Reserve in 1933, he served on the board of Army officers in November 1941 (while on inactive status) to determine the basis on which the War Department would work with OCD to organize and train CAP. The board recommended Blee be appointed as the aide to the CAP National Commander, and he was soon called to active duty by Curry and placed in charge of training and operations for CAP. Blee may not have been an expert in antisubmarine warfare, but with pilot ratings in airships and balloons and as an observer in airplanes, he understood civil aviation and possessed military experience and seniority to integrate the civilian aviators into a military command structure. He personally drafted and released practically all instructions, guidance, operational procedures, and assorted memoranda guiding CAP's coastal patrol effort.[4]

CAP's efforts required addressing questions of the legality of using civilians in military operations. As early as 21 January, Curry inquired with the judge advocate general, Maj Gen Myron C. Cramer, about the steps necessary to ensure CAP members would be considered lawful noncombatant belligerents treated as prisoners of war. Among Curry's queries were "is lettering 'U.S.' in lower segment of insignia sufficient to indicate the Federal nature of the organization?" and whether the size and location of the insignia worn on the shoulder were satisfactory.[5]

In acknowledgment of Curry's letter, Lt Col Archibald King, of the Judge Advocate General's department, provided an opinion on behalf of Cramer with answers to the CAP questions on 29 January. In accordance with the Annex to Hague Convention No. IV of 18 October 1907, of which the United States and Germany were signatories, CAP members qualified as lawful belligerents. To be entitled to belligerent status, individuals required a distinctive emblem, although not necessarily a uniform. King thought it advisable for the emblem to have some connection with the country for which it served and that "US" in the lower field of the emblem would be sufficient. For purposes of location, King considered the current CAP shoulder patch insufficient, with personnel better served by a larger breast insignia for easier identification. Accompanying the distinctive emblem, King recommended CAP aircraft be marked with the regular Air Corps markings, a distinctive CAP device, or both. King "deemed it advisable" for notice of the markings to be provided to the nation's enemies through Swiss diplomatic channels. Should the War Department approve King's responses to Curry's questions, he recommended reply be made "in harmony therewith."[6]

Curry's questions were grounded in military reality. U-boat commanders also had guidance that if an aircraft was sighted too late to submerge, then the boat would stay on the surface and fight the aircraft off with antiaircraft weapons.[7] Had a CAP aircraft been shot down and the crew captured, a diplomatic incident might have resulted. On 12 March 1942, Blee issued guidance for all CAP members to be "thoroughly familiar" with the War Department's rules of land warfare, which instructed personnel to provide only "name, grade, and serial number" in the event of capture.[8]

When the coastal patrols began that March, the members may not have appeared physically uniform, but their semimilitarization was well under way. Coastal patrol personnel wore CAP shoulder sleeve insignia featuring the letters "US"; by mid-July, CAP National Headquarters instructed every member to wear this same design.[9] CAP coastal patrol aircraft flew with roundels on the wings and fuselage identical to the shoulder sleeve insignia consisting of a red three-bladed propeller centered on a white triangle atop a blue disc; some aircraft roundels also included the letters "US" beneath the triangle, although these were soon ordered removed. Beginning on 1 August, CAP National Headquarters ordered the removal of the red propeller to distinguish the CAP aircraft on coastal patrol duty from those not assigned to the operation.[10] On 8 May Cramer issued an additional opinion in response to a Blee inquiry, declaring that CAP personnel on coastal patrol duty "are accompanying or serving with the Army of the United States in the field and that under the provisions of Article of War 2 (d) they are amenable to military discipline and subject to the jurisdiction of military courts."[11]

The first task forces at Atlantic City, Rehoboth, and West Palm Beach initially began operations with 59 personnel, with 15 pilots and observers respectively and up to 15 aircraft flying a daily average of 40 hours.[12] Pilots had to have a minimum of 200 flying hours, and pilots and observers required a practical working knowledge of air navigation. Aircraft assigned to coastal patrol duty had to be rated with 90-horsepower or greater engines and feature a two-way radiophone transmitter. All aircraft were required to be equipped for instrument flying with a minimum of a sensitive altimeter, airspeed indicator, compensated magnetic compass, tachometer, turn-and-bank and rate-of-climb indicators, and a clock with a sweep-second hand.[13]

Personnel on coastal patrol duty received a modest amount of financial support from the War Department. In May, CAP National Headquarters also published personnel per diem rates for coastal patrol task forces scaled to individual assignments. Base commanders received $10 daily. Pilots and pilot-observers received $8, while observers on nonpilot status received $7. At the low end of the pay scale, plotting board operators, clerk typists, apprentice mechanics, and servicemen received $5 in per diem.[14] These funds would theoretically cover costs ranging from uniforms and personal equipment to housing and meals. Aircraft owners and/or operators also had to carry accident, crash, passenger and public liability, and property damage insurance. Before any coastal patrol pilot took off on his first flight, a pilot had to pay his base commander the latter insurance premium.[15] CAP National Headquarters created a sliding scale of hourly reimbursement rates for coastal patrol aircraft, initially set from a low of $9.69/hour for 80–120 horsepower to a high of $41 for those with 400–445 horsepower engines. These funds would cover all expenses incidental to operation, maintenance, overhaul, repair, depreciation, replacement, and crash and accident insurance.[16]

The aircraft themselves, all prewar commercially produced models, represented a mix of nearly two dozen different airframes and 10 engines. In terms of aircraft models, the noninclusive list includes Aeronca, Beechcraft, Bellanca, Buhl, Cessna, Culver, Curtiss, Fairchild, Fleet, Fleetwings, Grumman, Harlow, Howard, Luscombe, Monocoupe, Piper, Rearwin, Ryan, Sikorsky, Stinson, Taylorcraft, and Waco.[17] Engine manufacturers included Continental, Franklin, Jacobs, Lambert, LeBlond/Ken-Royce, Lycoming, Pratt & Whitney/Wasp, Ranger, Warner, and Wright.[18] Over half of the aircraft on CAP coastal patrol service consisted of either the Stinson Voyager 10A or the Fairchild Model 24. The former featured a 90-horsepower Franklin engine, a 34-foot wingspan, and a cruising speed of 108 miles per hour at a range of 330 miles with room for two (or three with optional bench). The larger Fairchild had room for up to four, and those with CAP service used either the Warner 145-horsepower radial engine or the inline 145-horsepower Ranger engine. The Model 24 featured a 36-foot, 4-inch wingspan, a cruising speed of 103 miles per hour, and a range of 525 miles.[19]

Radio equipment located at the bases proved as valuable to CAP as the aircraft. Bases required an estimated outlay of at least $3,000 to provide radio transmitting and receiving equipment. Operations could grind to a complete halt without functional radios, with CAP National Headquarters explaining that the patrol's purposes would be "essentially defeated before it starts" without adequate radio communications.[20] While some states provided funding to buy aircraft radios, acquisition of larger ground units and construction of radio towers often fell to the ingenuity of the base personnel to borrow and improvise.[21] Individual aircraft had to have working radiophone transmitters of at least 7.5 watts power on 3105 kilocycles and a radio receiver able to receive in the 200–400 kilocycle airways band. They also required a one-quarter-wave Hertz trailing-type antenna with 75 feet of copper aerial to accompany the radiophone transmitters. Each CAP coastal patrol base had to possess: one ground transmitter of at least 15 watts' power, able to transmit in the airways band; one (but preferably two) radio ground receivers to receive aircraft radiophone signals; two radio operators licensed by the Federal Communications Commission; two radio mechanics to maintain the aircraft and ground equipment; and its own frequency and call sign, operated under certificates of authorization issued by the Chief Signal Officer of the Army.[22]

Figure 9. Radio testing and maintenance performed at Coastal Patrol Base No. 11, Pascagoula, Mississippi. (Photograph courtesy of the Morse Center.)

Unlike the motley array of aircraft and communications equipment, the personnel at the task forces organized themselves along the prevailing national gender and racial lines. Curry publicly aligned with OCD policy and proclaimed CAP overall would not discriminate regarding race or sex. The coastal patrol effort, however, did not officially adhere to this policy. CAP coastal patrol aircrews were exclusively white males. Women could work as radio operators, administrative section heads, plotting board operators, or clerk typists at the task forces but were prohibited from coastal patrol flying.[23] When Bettie Thompson, secretary treasurer of the Middle Eastern Section of the Ninety-Nines, asked Blee to provide an explanation as to why women could not fly as pilots or observers on coastal patrol missions, the colonel avoided specifics.[24] Instead, he explained CAP's desire to use women "in everything possible" but that "we cannot assign them to the performance of flying missions in the Theater of Operations."[25] Johnson, in response to a query from Arlene Davis of Lakewood, Ohio, explained "the Army does not want women on the coastal patrol," adding, "obviously there are dangers involved and if a plane falls in the water, a man has a better chance of getting out."[26] Sexism aside, several women at the first two task forces managed a few flights as observers prior to CAP National Headquarters limiting flights to men only in early April.[27]

The women at the bases shared the same enthusiasm for aviation as the men and served the country before themselves. Academy Award winner Mary Astor served as a radio operator at Base 12 in Brownsville, Texas, in the summer of 1942, drawn to CAP by her interest in flying. In a letter to CAP National Headquarters, she lamented not having more time to serve on active duty but hoped to return at some future date.[28] For those aircrew confronting an in-flight emergency, the voice of a female radio operator helped calm the men's nerves.[29] At almost every coastal patrol base, wives of other base personnel worked in the operations and administrative offices. Couples young and old worked side by side, and some brought their children with them for the duration of the operation.

CAP National Headquarters never explained why there were no African-American pilots or observers. As early as 17 December 1941, Cornelius R. Coffey, vice president and founder of the National Airmen's Association of America, wired Curry that "Negro air pilots throughout the United States are anxious to serve this country in all branches of the air service. Please enlist us in the Civil Air Patrol and command us as you see fit."[30] Curry replied the following day and noted no restrictions on CAP membership as to race, creed, color, or sex, explaining that "ability to do [the] job [is the] only consideration beyond patriotism."[31] On 20 March 1942, Jack Vilas, commander of the Illinois Wing, swore in Coffey as commander of the 111th Flight Squadron in Chicago with his business partner and the nation's first licensed African American female pilot, Willa Brown, as squadron adjutant. "Civil Air Patrol Does the 'Impossible' in Illinois" read the headline in the *Chicago Defender*, as the 111th, with its 25 black and white, male and female pilots, became the first racial and gender integrated, uniformed operational flying unit in American history.[32] This civil rights achievement remained an unsung wartime milestone, having received no mention in any press release by CAP National Headquarters, OCD, or the Army.[33]

As with the coastal patrol women, African Americans were confined to the ground. The men worked in aircraft maintenance or addressed airfield facility issues. The women predominately engaged in cooking or janitorial work. PFC Oliver Hamilton of the engineering department at Fifth Task Force, Flagler Beach, Florida, is seemingly the only African American even mentioned by name in a CAP coastal patrol history.[34] Although the number of African American male pilots was small, interest in active duty service remained substantial. In September 1942, men from Coffey's squadron flew a 2,000 mile cross-country trip to gain experience—and presumably demonstrate competence in the face of prejudice—so "that they might participate in some of the more daring missions" of CAP.[35] Whatever the case, the era's unwritten prejudices and the machinations of Jim Crow ensured CAP coastal patrol had a monochromatic racial identity.[36] Surviving records of African American personnel at the CAP coastal patrol bases remains limited to photographic evidence of nameless faces.[37]

The first CAP coastal patrols in March were almost entirely improvised affairs—trial and error.[38] Through reports, correspondence, and operational experience, procedures and profiles for coastal patrol missions emerged. Patrols consisted of loose formations of two aircraft with two-man crews (pilots and observers), flying for several hours at a time up to 15 miles offshore at altitudes from a few hundred to a thousand feet.[39] Aircrews used paper and pencil for navigation, though when possible they relied on dead reckoning with known navigation fixes or waypoints such as lightships, buoys, or shipwrecks.[40] Additional navigational instruction for aircrews came from among task force personnel or from civilian aviation experts assisting CAP.[41]

Figure 10. A standard two-ship formation of aircraft of the Second Task Force, Rehoboth, Delaware, on patrol off the adjacent coast. (Photograph courtesy of Henry E. Phipps via the Morse Center.)

CAP task forces turned to the Army Air Forces for antisubmarine warfare training. The Army, however, admittedly began the war with little guidance to provide its own aircrews. CAP received some training materials from I Air Support Command for familiarization with U-boat tactics to help improve spotting accuracy.[42] Other training materials from the Navy were distributed to CAP coastal patrol personnel over the course of the year.[43] In October 1942, the Army stipulated that training for CAP coastal patrol personnel was the function and responsibility of CAP National Headquarters. Not until late May 1943 would antisubmarine warfare training courses taught by Army instructors be offered to CAP personnel. A select aircrew from each base attended the "Familiarization in Antisubmarine Warfare Technique" program in Atlantic City before returning to their home bases to impart the new knowledge on their colleagues.[44] Until then, training varied by base. To refine bomb aiming and dropping, several bases created targets consisting of a submarine outline with something akin to a conning tower in the middle. These would allow aircrews to perfect their approaches and dive angles while dropping practice bombs.[45]

America's military and CAP faced an intense learning curve in antisubmarine warfare while confronting an opponent with over two years of operational experience. Despite increases in operational forces, U-boats continued to attack and sink ships. The East Coast shipping losses in April 1942 varied little from those of March. The Eastern Sea Frontier war diary begrudgingly acknowledged "in the submarine warfare, April was almost an exact repetition of the preceding month" and recorded the Eastern Sea Frontier as "the most dangerous area for merchant shipping in the entire world."[46] Andrews asked King for additional aircraft and ships to combat the U-boat menace. He argued that convoying must be adopted and intended to implement a convoy system in mid-May if he could acquire more escorts.[47] Andrews warned that until he had more planes and warships at his disposal, losses would remain "extremely critical."[48] In the interim, he began implementing a partial coastal convoy system deemed the "Bucket Brigade." Merchant ships now zigzagged by day with aerial coverage and anchored at night inside protected bays or anchorages guarded by sea mines, antisubmarine nets, and defensive patrols, spaced 120 miles apart along the Atlantic Coast.[49]

With three task forces established, CAP attempted a modest expansion. In mid-April, CAP planned a rollout of bases in Florida, New York, and Virginia so eight would be operational by 22 April. Each base had an estimated monthly cost of $20,000.[50] The rollout coincided with a change in leadership at CAP National Headquarters. On 17 March, Curry received a new assignment as commander of the Fourth District, Air Forces Technical Training Command in Colorado.[51] James Landis named CAP's executive officer, Earle L. Johnson, as Curry's successor. A Massachusetts native, Johnson grew up in Ohio and graduated from Ohio State University, where he played right guard for the Buckeye football team. In 1926, he was elected to the Ohio House of Representatives along with his friend David S. Ingalls, the Navy's first and only fighter ace of World War I. Ingalls cultivated and refined Johnson's interest in aviation, and he soon became a passionate promoter of civil aviation. Named director of the Ohio Bureau of Aeronautics in 1939, Johnson founded and began organizing the Ohio Wing of the CADS in September 1941. On Christmas Eve, he joined CAP National Headquarters as assistant executive officer under Gill Robb Wilson. Receiving a presidential commission as a captain in the Army Air Forces on 1 April 1942, Johnson remained essentially a civilian in uniform, the bridge between the CAP volunteers and the War Department.⁵²

Figure 11. Col Earle L. Johnson, as CAP National Commander in 1946. (Photograph courtesy of the Morse Center.)

Shortly after being named national commander, Johnson traveled to the First Task Force at Atlantic City. There he looked up one of the base pilots, Manhattan stockbroker and financier Isaac W. "Tubby" Burnham II, and asked if he would be interested in opening a fourth CAP task force. The base in question would provide coverage between Cape May, New Jersey, and Norfolk, Virginia, following the movement of German U-boats southward to less patrolled waters. Burnham agreed to take on the challenge of establishing the base from the ground up. On 15 April, Johnson ordered Burnham to fly to Parksley, Virginia, to advise him if the field was sufficient for operations. Flying down from Atlantic City in his Stinson Voyager 10A, Burnham finally deduced that a barely discernable outline of runways located at a chicken farm a mile southwest of the town represented the airport. The airport was small, bounded on three sides by deep ditches and trees, and bisected by a busy road. With the support of the town leadership and optimistic of success, Burnham told Johnson a base could operate at the location. The Fourth Task Force activated the next day, but immense work remained to turn a chicken farm into a functional airport.⁵³

Figure 12. Aerial view of the field of the Fourth Task Force, Parksley, Virginia. The field, visible to the left of the town, was carved out of a former chicken farm. (Photograph courtesy of William G. Bell via the Morse Center.)

CAP's movement to Parksley pushed the limitations of the volunteers forced to resurrect an airport out of farmland. Since they lacked adequate federal funding, Sun Oil combined its original $10,000 with some $5,000 apiece from eight oil companies to create a Tanker Protection Fund on 10 April to help stand up Parksley and assist other bases with the purchase of life preservers, engines, radios, and other equipment needs. The Tanker Protection Funds also helped tide over the First and Second Task Forces when promised federal monies failed to materialize in a timely fashion. The exact totals and the identities of the oil companies that donated funds remain unknown, but the amount added up to between $40,000 and $45,000. By early 1944, CAP National Headquarters had expended all but approximately $5,000; the monies helped quite a few commanders like Burnham get aircraft safely out on patrol.⁵⁴

Additional assistance came from state government. The Virginia General Assembly authorized a $5,400 appropriation for the Virginia

Wing that provided two-way radios for base aircraft. The state road commission and Works Projects Administration brought in equipment to extend the grass runways, fill ditches, and clear trees. In addition to building the physical airstrip, Burnham and CAP learned with Parksley that efforts to acquire aircraft, radio equipment, and personnel exclusively from individual state wings would not always be sufficient to meet operational needs. The base consequently featured personnel and aircraft drawn from more than 10 states and the District of Columbia. This practice of bringing in aircraft and personnel from across the country thereafter became routine at Parksley and for future CAP bases. Parksley's first patrols lifted off 17 May, a month after activation.[55]

Burnham's challenges in transforming a disused civil airfield-turned-chicken farm into a semimilitary air base were repeated over the next few months along the Atlantic and Gulf coasts. CAP base commanders faced rather daunting tasks, outlined by CAP National Headquarters in a two-page document on necessary procedures for establishing coastal patrol bases. Bases could cover from 80 to 100 miles of shoreline, with aircraft flying a cumulative average of 25 to 30 hours daily. Ideally the base would be centrally located within the assigned patrol area and have available shop and hangar sheltered areas. In addition to the space necessary for aircraft storage and maintenance, bases had to have room for the radio equipment and plotting board. An operations room, a pilot's lounge, and an administration office should be available. If space were not available for the latter, the administration office could be housed elsewhere. Commanders needed to plan to quarter all personnel together at the same place. This would usually consist of a boarding house or small hotel in the nearby community if not adjacent to the field. If away from the field, commanders needed to arrange transportation. National Headquarters suggested seeing if local towns had any voluntary transportation corps to provide the necessary transport. If not, commanders needed a minimum of three cars or stations wagons, usually provided by the members themselves.

Planes and personnel remained the bread and butter of the bases. In addition to having 12 to 15 aircraft with matching numbers of pilots and observers, commanders needed to ensure their smooth operation. Each pilot and observer required procurement of one safety vest and one life buoy or similar equipment. Whenever possible, commanders needed to obtain smoke bombs or small flare pistols.

"The more safety precautions and equipment the better," and headquarters recommended every base have at least one amphibian/flying boat on duty.[56]

To keep the men and aircraft aloft, bases required a viable maintenance section. Every base required an engineering section with a supervisor, the necessary number of mechanics to properly maintain the aircraft, and two or three mechanic's helpers and airdrome helpers. Commanders needed to designate someone within the engineering section to manage supply matters. To ensure funding came to the bases, an office section stood up to handle the operations reports and the various pay vouchers.

An operations section had to be established. This would include an operations officer, an intelligence officer, assistants for both, and such messengers and other helpers as needed. Adjacent to operations was the radio section, with at least two radio operators and an aircraft radio mechanic. Within operations, headquarters recommended having someone constantly on duty on the plotting board to keep track of all aircraft movements and plot future positions.

Once commanders had all of these matters resolved, they needed to address training. CAP National Headquarters would not permit any pilot, observer, or airplane to fly an active duty mission until properly covered with published insurance requirements. After the preliminary work had been done and the designated Army liaison officer received guidance from his higher headquarters, the CAP base commander needed to meet with his Army counterpart to understand each other's areas of responsibility. Telephone and teletype communications between the CAP base and the Army needed to be installed and tested. After this, the assigned Army intelligence officer needed to brief the CAP base's pilots, observers, radio operators, plotting board personnel, and intelligence officers so they would have a clear understanding of their duties.

Only after commanders completed these tasks could the base commence flying test patrol missions. National Headquarters recommended at least two days of these test patrols. Aircrew needed to become completely familiar with their patrol territory and receive a thorough foundation in the proper method of making good a course out of sight of land, making landfalls, and returning to emergency fields in simulated bad weather conditions. Only then, when the base commander was confident in everyone's abilities, should official patrol missions get a green light. Commanders were instructed to immediately replace CAP base personnel who failed to demonstrate qualifications or failed in their duty. "It is extremely dangerous to keep anyone on duty at a base who is not thoroughly competent," concluded guidance to base commanders.[57]

Coinciding with Burnham's arrival at Parksley, CAP National Headquarters began releasing guidance on the operation and organization of the coastal patrol bases. Beginning on 15 April, all base commanders were directly appointed by Johnson and operated directly under Blee at national headquarters.[58] Blee issued a coastal patrol base table of organization the following day for 15 aircraft and 59 personnel.[59] The task forces operated as small versions of Army Air Forces bases. The base commander had to be an active or former pilot; operations and assistant operations officers had to be pilots meeting the same criteria as all other assigned pilots or pilot-observers. CAP National Headquarters preferred that the other base staff officers (intelligence, engineering, assistant engineering, and airdrome) hold pilot certifications. Aircraft and personnel assigned would have to serve a minimum of 30 consecutive days, although national headquarters preferred those able to serve at least 90 consecutive days.[60]

In the broader context of the war, the situation along the nation's coast was shifting in favor of the Americans. Army Air Forces units increasingly carried radar equipment and had begun to record actual attacks on U-boats.[61] Despite the heavy shipping losses in April, that same month Andrews could field a more formidable opposing force. He now controlled over 100 surface ships, 100 Navy and Coast Guard aircraft, and four blimps, plus 100 Army aircraft and the Atlantic Fleet's Catalina flying boats, not to mention the use of 23 destroyers that month. At mid-month the Navy claimed its first U-boat kill (*U-85*), and the Germans recognized that the hunting grounds around Cape Hatteras were too well defended.[62] Andrews's Bucket Brigade and other defensive additions at last made American antisubmarine measures felt by the enemy.

After a tour of Army units at Morrison Field, Lovett stopped in to inspect the equipment of CAP's Third Task Force. Fortuitously, he visited the one coastal patrol unit then in operation with the strongest semimilitary bearing. Vermilya's drilled, uniformed volunteers and aircraft exuded a professionalism beyond any perception of civilian amateurism. The Third Task Force made a strong impression on Lovett, and he wrote to Arnold inquiring about using the unit for

CAP coastal patrol work.[63] Arnold forwarded Lovett's message to Major General Bradley at First Air Force headquarters, who acknowledged that the Third Task Force "has performed commendable service conducting patrols" that permitted the movement of Regular Army units to reinforce patrols in the Jacksonville area.[64]

As the balance of power shifted with some promising success against the U-boat threat, the veil of secrecy began to lift on the coastal patrol operation. In the wake of Lovett's memorandum to Arnold, Bradley, on 20 May, publicly disclosed the existence of CAP's coastal patrol operation.[65] His announcement coincidentally came days after the destruction of *U-85* by the destroyer *Roper* (DD-147) off North Carolina, the first U-boat sunk off the East Coast.[66] The press release about the CAP task forces explained how the members "brought their own planes, tools and spare aircraft parts with them and they arranged to secure sufficient radio equipment" to communicate with patrol planes. "The flying minute men have established strict military discipline at the bases on their own initiative," continued the release, and CAP leaders had organized classes in "Morse Code, signaling, infantry drill, and similar subjects" at the task forces and at units nationwide. In a particularly eye-opening statement of praise and questionable credibility, Bradley declared "several sinkings scored by Army and Navy bombers are credited directly to a tip-off by civilian volunteers assigned to patrol work in certain Atlantic areas off the United States."[67]

Although the claims of destroyed U-boats attributed by Bradley were more fabrication than fact, his message struck a positive chord with CAP as a public acknowledgment of the Army Air Forces' appreciation for the civilian effort. One week later, the 27 April issue of *Life* featured a photographic essay of the experimental operations at Atlantic City, providing the public with the first glimpses of civilian aircraft patrolling the shipping lanes. The article featured photos of tankers at sea, air base operations, and aircrews running to their motley fleet of planes.[68]

The national press announcements also represented an acknowledgment of the efforts of CAP National Headquarters to restrict any disclosure of information about the initial task forces. Throughout the existence of the coastal patrol operation, CAP National Headquarters instructed personnel to neither take cameras into the air nor report anything about the effort.[69] Even before the coastal patrol experiment commenced, CAP's weekly newsletter cautioned about the need for secrecy because as the patrol "goes more and more on military missions, confidences must be strictly observed."[70] As the CAP coastal patrol units received classified information from the Army and Navy, all loose talk was discouraged, and National Headquarters instructed all unit intelligence officers to keep the coastal patrol story out of the papers.[71] The news stories released in April 1942 represented the majority of public statements about the coastal patrol effort for the entire year, although CAP leadership remained vigilant to squash national stories even if newspapers near the coastal patrol bases made CAP's operations the local region's worst-kept secret.[72]

Consistent funding for the CAP effort emerged by mid-1942. CAP National Headquarters previously estimated each task force could maintain 12–15 operational aircraft for approximately $20,000 per month, and they retained this figure as the overall effort expanded.[73] In May, the War Department transferred $160,000 to OCD to establish the Fifth through Eighth Task Forces in Florida and Georgia.[74] Through late June and early July, OCD director James Landis made a favorable impression on congressional appropriations committees when testifying about the CAP coastal patrol effort. Arnold provided support with his testimony, explaining the Army's desire to expand CAP operations for fiscal year 1943.[75] Congress approved the funds, and on 4 July Arnold authorized the War Department Budget Section to further fund CAP operations through lump sum transfers to OCD.[76]

At the end of April 1942, the adjutant general, Maj Gen James A. Ulio, requested that Arnold place certain elements of the CAP under War Department jurisdiction for use in frontier defense and for purposes where Army aircraft and combat crews could be replaced. Ulio suggested incorporating CAP members into the Army Specialist Corps and directed Gen Brehon B. Somervell, commanding general, Army Service Forces, to provide the necessary funds for this project.[77] Established on 26 February, the corps intended to bring into the War Department skilled civilians who possessed particular professional, technical, or scientific qualifications to enable them to perform military duties and free uniformed personnel for combat and command duties.[78]

Maj Gen Millard F. Harmon recommended against Ulio's suggestion. As Harmon explained to the War Department, the immediate need for coastal patrol assistance negated the time and administrative requirements of turning civilians into military personnel. As CAP members flew and maintained their own aircraft, they could operate until the civilian aircraft wore out, at which point the CAP personnel would not be an Army responsibility. But if CAP members transferred to the Specialist Corps, then they would expect to fly Army aircraft, requiring both training and military equipment. Retaining the volunteer esprit de corps on a reimbursable basis thus gave the Army a more flexible, affordable resource.[79]

Blee also weighed in on the issue of CAP's organizational status as an OCD component serving the War Department. In his opinion, leaving CAP as an auxiliary of the Army Air Forces was the most prudent option as the organization had already passed through periods of mobilization, organization, and instruction before rapidly expanding. "Now is no time to break up the Civil Air Patrol and start building up another organization to the same job," he concluded.[80] On 4 July, Brig Gen Laurence S. Kuter, deputy chief of staff, Army Air Forces, informed the War Department budget officer that General Arnold authorized the continued utilization of "the services of the Civil Air Patrol for an indefinite period of time and to extend the scope of operations to include missions other than anti-submarine patrol and to extend the operations area to other than the East Coast."[81]

Events in May signaled a significant shift in the Battle of the Atlantic for both CAP and the Navy. The first escorted convoy sailed south from Hampton Roads, Virginia, on 14 May, while a northbound escorted convoy sailed from Key West, Florida, the following day.[82] Three days before the convoy sailed, Blee issued instructions to use the CAP task forces, clarifying that aircraft would engage in patrol and "convoy service" as assigned from the respective sea frontiers.[83] The convoy system along the Eastern Seaboard witnessed an almost immediate reduction in losses. With easy targets no longer available on the East Coast, German Vice Admiral Doenitz shifted his submarine operations southward along the Florida coasts, the Caribbean, and into the Gulf of Mexico where aerial defenses were in short supply.[84]

This southward movement of the U-boat offensive directly contributed to the arming of CAP coastal patrol aircraft. At the beginning of May, three U-boats—*U-109*, *U-333*, and *U-564*—plied the waters off Florida near Morrison Field. With multiple submarines operational, CAP coastal patrol aircraft found themselves overflying an active battlespace. On 1 May, dawn patrols from the Third Task Force sighted the British motor merchant *La Paz* afloat with a torpedo hole in her side, courtesy of *U-109*. Two days later, a CAP patrol located the British steam merchant *Ocean Venus* sinking rapidly with multiple survivors in rafts and boats nearby from a morning torpedo strike by *U-564*. Another patrol located the Dutch steam merchant *Laertes*, torpedoed and abandoned, victim of an attack by *U-109* that killed 18 crewmembers.[85]

At noon on 4 May the next day, Rex Bassett overflew the British steam tanker *Eclipse* while ferrying an aircraft from Miami to Morrison Field. Minutes after the small CAP aircraft passed the tanker, a torpedo from *U-564* struck the ship, which soon settled by the stern onto the shallow bottom. Upon landing at Morrison Field, Bassett incredulously received the sinking report. Patrols from the base flying eight miles south of the incident turned north and searched unsuccessfully for the submarine. Later that day, three base aircraft reported sighting a submerged submarine, possibly *U-564*, stalking another tanker. The aircraft dove on a periscope and the submarine appeared to break off the attack. Col Dache M. Reeves, commander, Headquarters, I Ground Air Support Command, sent commendations to the CAP aircrews.[86]

This small measure of success proved a distant memory for the dawn patrols of 5 May. CAP aircraft came upon the American steam merchant *Delisle* partially submerged with a torpedo hole in her side. Despite intense searching the enemy below remained hidden and the CAP men made no contact reports. Later that day, the Third Task Force received new orders to extend patrol operations to Daytona Beach, 190 miles north of the base. That evening, planes flew up to the airfield at Melbourne, positioned midway between West Palm Beach and Daytona Beach, to cover the larger patrol area more easily. As the CAP personnel rested for the night, in the early hours of 6 May, *U-333* managed to torpedo the American tankers *Java Arrow* and *Halsey* as well as the Dutch freighter *Amazone*, damaging the former and sinking the latter two.[87]

At dawn on 6 May, a patrol out of Melbourne flown by Carl N. Dahlberg with observer Earl Adams spotted a partially submerged submarine, most likely *U-109*, stalking a tanker. Dahlberg turned his aircraft around and dove on the U-boat, which submerged. He called in a Navy patrol craft, which dropped several depth charges near the location. Reeves commended Dahlberg and Adams for saving the tanker.[88] Around dusk that evening while returning to Melbourne, an aircraft piloted by Marshall E. "Doc" Rinker with Thomas C. Manning as his observer reported sighting a U-boat just off Cape Canaveral "in such shallow water that the U-boat rammed its prow into the mud bottom while attempting to escape." The CAP aircrew radioed for help, but an armed aircraft dispatched from nearby Naval Air Station Banana River did not arrive until well after the boat had vanished.[89] The U-boat in question may have been *U-109*, which reported being near Bethel Shoal, east of Vero Beach and south of Cape Canaveral, Florida, intent on closing in on and attacking a tanker in shallow water but which broke off attack due to a circling aircraft. To the untrained eye, the maneuvering boat in shallow water may have appeared to have run aground.[90]

The seventh of May brought relative quiet beyond two false reports. But on the eighth, *U-564* torpedoed and sank the American steam merchant *Ohioan*, killing 15 crewmembers. By 9 May, the three U-boats operating near the Third Task Force had sunk six merchantmen and damaged three others without incurring any damage.[91] Around noon on 14 May, CAP aircraft from Morrison Field overflew the drifting hulk of the neutral Mexican tanker *Portrero Del Llano*, torpedoed earlier in the morning off Miami by *U-564*. John A. Keil, Third Task Force intelligence officer, photographed the wreckage and forwarded the photographs to Johnson at CAP National Headquarters. Keil later learned Johnson wired the images to the Mexican government at the request of Major General Harmon. After *U-106* torpedoed a second neutral tanker *Faja de Oro* on 21 May, Mexico declared war on Germany the following day.[92]

The experiences of the Third Task Force would cause a notable policy shift in CAP coastal patrol operations. The unidentified submarine grounding incident of 6 May, coupled with the increased U-boat activity in the Palm Beach area, caused a stir in Washington. Assistant Secretary of the Navy Ralph Bard wrote to Vice Adm Frederick J. Horne, vice chief of naval operations, and reported a phone call he received from a "very responsible man" reporting the attacks off Palm Beach. He stated that no Navy ships assisted the survivors while the Army bombers at Morrison Field had "no bombs, and no authority to do anything but reconnaissance."[93] Arnold passed Bard's letter to Marshall and mentioned his receiving a report of a submarine seen "in such shallow water that it required some 20 to 25 minutes to get clear. All this time one of our small reconnaissance planes was yelling for help while it circled above." In reply, Bradley at First Air Force stated the destruction of submarines remained a Navy matter.[94]

Figure 13. British tanker *Eclipse* torpedoed on 4 May 1942 off Boynton Beach, Florida. Survivors of the attack are seen in the foreground on the beach. (Photograph courtesy of the Morse Center.)

Arnold decided Army aviation needed to strike back. He wrote Marshall, suggesting that all Army Air Forces units on antisubmarine activity be placed under the immediate control and authority of the commanding generals of the defense commands and to arm all small reconnaissance aircraft, also with 100-pound bombs, although he mentioned "no bombsights will be used."[95]

Arnold's request to arm CAP aircraft had previously been explored. Reed Landis had suggested to Curry as early as 13 February that CAP needed to inquire with the CAA about the rules or requirements for CAP members to carry arms when operating under military authority on civil aircraft.[96] On 5 May the CAA's general counsel concluded carrying explosives in civil aircraft was prohibited, except while in possession of the armed forces.[97] Legally speaking, Arnold's order could proceed. He telegraphed Bradley, ordering First Air Force to "equip the Civil Air Patrol airplanes operating under the First Air Support Command with one hundred pound bombs for use against submarines."[98]

Figure 14. Oil slick and debris from the merchant steamer *Ohioan*, torpedoed and sunk by *U-564* on 8 May 1942 off Boynton Beach, Florida, with a loss of 15 dead. (Photograph courtesy of the Morse Center.)

On 11 May, Brigadier General Kuter directed that all "puddle jumpers" on antisubmarine patrol—ergo, CAP aircraft—be modified to carry and release 100-pound bombs.[99] Five days later, I Ground Air Support Command included language in its letters of instructions for the Fifth through Eighth Task Forces that read, "Airplanes of the C.A.P. units when equipped with suitable racks *are authorized to carry and drop bombs*" (emphasis added). Pilots would receive instruction in handling and dropping procedures, and no unit would carry or drop any live ordnance until specifically directed by I Ground Air Support Command or an authorized representative.[100] On 20 May, Arnold received a memorandum from Maj Gen Dwight D. Eisenhower, army assistant chief of staff for operations, providing the Air Forces chief with additional supporting authority and requesting immediate steps to equip all planes with racks to carry bombs and/or depth charges and bombsights where possible. Eisenhower added that "it is understood that C.A.P. planes which carry bombs cannot be insured. Have necessary action taken which will protect the owners of the planes in event of their damage or loss."[101] By late June, I Ground Air Support Command updated all the task force mission statements for CAP patrols "to take all action within their means to destroy any enemy sighted."[102] In mid-July, CAP National Headquarters increased the hourly insurance reimbursement rates for aircraft equipped with bomb racks, particularly with a $3 addition for insurance against liability for explosion.[103]

In due course, the CAP coastal patrols received deliveries of ordnance and added new personnel. The nearest Army Air Forces base received orders to assign a detail of four enlisted ordnance personnel (one noncommissioned officer, one corporal, and two privates) along with one bomb service truck and accessories and one bomb trailer to each coastal patrol base. These men would handle the arming and disarming of all CAP aircraft with either practice bombs, 100-pound demolition, or 250- or 325-pound depth bombs. Base commanders now included bomb dumps on their list of responsibilities and accountability for reporting expenditures in training and patrols. CAP aircrews had authority to drop ordnance and received instruction on how to remove and install safety pins in the bomb fuses. All handling of live ordnance was otherwise handled only by Army Air Forces personnel.[104]

Figure 15. A CAP Stinson Voyager 10A armed with an AN-M30 100-pound, general-purpose demolition bomb at the Fifth Task Force, Daytona Beach, Florida, 24 June 1942. The aircraft is believed to be of the Third Task Force, Lantana, Florida. (Photograph courtesy of the Morse Center.)

The Army's decision to arm and empower CAP aircrews demonstrated a growing trust in the civilian organization and a sense of urgency in staunching the losses in the Battle of the Atlantic. CAP subsequently issued corrective actions to increase its military character to measure up to its new responsibilities. All volunteers serving on coastal patrol task forces were required to execute an active duty oath, a variant on the same oath executed by military officers. The last paragraph of the oath conformed to CAA wartime policy, reading: "In the event that I shall not report or be available for active duty at any time during said term or any extension thereof which I shall perform all duties assigned to me, I hereby consent to the revocation and cancellation of my license to own, operate and service any aviation and radio equipment." Base personnel received instruction to further devote one hour per week to infantry drill "in order to develop precision of action, general efficiency and esprit de corps."[105]

In late May, CAP National Headquarters also instructed all task forces to organize an armed guard to patrol the coastal patrol bases to protect buildings and property. As the military owned the demolition and depth bombs used by CAP, responsibility for the safety of this equipment necessitated placing guards at the bases. Guards, some as young as 16 years old, would be on duty 24 hours a day, seven days a week operating three 8-hour shifts. They received a per diem of $5 and had to furnish their own 12-gauge shotguns, whistles, and flashlights, although ammunition would be supplied by the Army.[106] The addition of 16 guards to each base represented the largest shift to the coastal patrol task force table of organization. By late August, CAP National Headquarters capped the maximum number of personnel at each base at 78, a numerical acknowledgment of the need for more volunteers to handle the increased tempo and complexity of operations.[107]

Now expanded in bases and personnel, CAP faced the challenge of conducting and sustaining patrol operations with the added responsibility of carrying armament and maintaining both the personnel and aircraft bearing the explosive burden. CAP's contributions to the war would thus be for the duration.

Notes

Epigraph. Burnham, "History of CAP Coastal Patrol No. 4," 8.

---

1. Morison, *Battle of the Atlantic*, 278.
2. US Air Force, *Historical Study No. 107*, 34; and Coates, *U-Boat Commander's Handbook*, 24–25, 78–81.
3. E. R. Baker to L. C. Martin, memorandum, subject: Employment of the Civil Air Patrol by the Army and the Navy in the Inshore Coastal Patrol on the Atlantic Coast, 8 April 1942, Folder "OCD Civil Air Patrol," Box 10, Entry 107A, RG51, NARA.
4. John F. Curry to Adjutant General, memorandum, subject: Report of Entry on Active Duty, 20 January 1942; Biographical Sketch of Harry H. Blee, Colonel, Air Corps, Training and Operations Officer, Civil Air Patrol, undated, BLS; War Department, Headquarters Army Air Forces, Special Orders no. 53, 4 November 1941; George E. Stratemeyer to Chief of the Army Air Forces, memorandum, subject: Civil Air Patrol, 8 November 1941; Report of Proceedings of Board of Officers, Headquarters Army Air Forces, 8 November 1941, CAP-NAHC; and War Department, Headquarters of the Army Air Forces, Dudley M. Outcalt to Air Inspector, memorandum, subject: Survey of the Civil Air Patrol, 8 March 1944, 5, Folder 3, Box 5, ELJ, WRHS.
5. John F. Curry to Myron C. Cramer, memorandum, subject: Status of Civil Air Patrol Members as Prisoners of War, 21 January 1942, Binder "Legal Status, Administrative Concepts, and Relationships of the Civil Air Patrol, 1941 to 1949," CAP-NAHC.
6. Archibald King to Myron C. Cramer, memorandum, subject: 1st Indorsement, 29 January 1942, Binder "CAP Uniform and Insignia Research, General Historical Files Vol II, Items 6–14," CAP Historical Foundation Collection (CHF), CAP-NAHC; and *Hearings before a Subcommittee of Finance on War Injury and Death Benefits for Civilians*, S. 450, 77th Cong., 1st sess., 12 and 15 March 1943, 91–92, 96–100.
7. Coates, *U-Boat Commander's Handbook*, 81.
8. OCD, CAPNHQ, Operations Directive No. 7, Rules of Land Warfare, 12 March 1942, Reel 38907, AFHRA. Blee's directive included extracts from War Department Basic Field Manual FM 27-10, Rules of Land Warfare.
9. OCD, CAPNHQ, Earle L. Johnson to All Unit Commanders, memorandum, subject: Uniform, Insignia, and Rank (GM-45), 17 July 1942, Binder "Civil Air Patrol–Establishment of Charts, Staff, General Memoranda, Training Memoranda," Box 2, Entry 205, Processed Documents Issued by the OCD–CAP, RG171, NARA.
10. OCD, CAPNHQ, Operations Directive No. 2, Display of Civil Air Patrol Insignia on Aircraft, 12 February 1942; OCD, CAPNHQ, Harry H. Blee to All Unit Commanders, memorandum, subject: Operations Memorandum No. 4, Operations Directive No. 2–Display of Civil Air Patrol Insignia on Aircraft, 14 April 1942; OCD, CAPNHQ, Operations Directive No. 23, Changes No. 2, Task Forces on Coastal Patrol Duty, 31 July 1942, A. William Schell Collection, (AWS), CAP-NAHC; and Robert H. Heartwell to Earle Johnson, 27 July 1942, Reel 38919. AFHRA.
11. Myron C. Cramer to Harry H. Blee, memorandum, subject: Disciplinary Powers of Army over Pilots of the Civil Air Patrol, 8 May 1943; and OCD, CAPNHQ, Harry H. Blee to All CAP Coastal Patrol Commanders, memorandum, subject: Military Status of CAP Coastal Patrol, 30 March 1943, BLS, CAP-NAHC.
12. OCD, CAPNHQ, Harry H. Blee to All Regional, Wing, and Base Commanders, memorandum, subject: Confidential Letter of Instruction No. 2, Organization of Coastal Patrol Bases, 16 April 1942, Reel 38907, AFHRA; and Robert Taylor to Robert M. Webster, memorandum, 22 April 1942, Reel 38918, AFHRA.

13. OCD, CAPNHQ, Confidential Letter of Instruction No. 2, "Organization of Coastal Patrol Bases," 16 April 1942, Reel 38907, AFHRA; and OCD, CAPNHQ, Harry H. Blee, Operations Directive No. 23-A, CAP Coastal Patrols, 26 August 1942, AWS, CAP-NAHC. With the addition of bomb racks and ordnance, horsepower ratings climbed from a minimum of 80 to a minimum of 90 in July. Earle L. Johnson to Bertrand Rhine, 29 May 1942, Reel 38920, AFHRA; and OCD, CAPNHQ, Harry H. Blee, Operations Directive No. 23, Change No. 1, Task Forces on Coastal Patrol Duty, 22 July 1942, Reel 38919, AFHRA.

14. OCD, CAPNHQ, Operations Directive No. 13, Reimbursement Schedules for Coastal Patrol Bases, 1 May 1942; and OCD, CAPNHQ, Operations Directive No. 13A, Reimbursement Schedules for Coastal Patrol Missions, 28 May 1942, AWS, CAP-NAHC.

15. OCD, CAPNHQ, Henry A. Hawgood to All Unit Commanders, memorandum, subject: General Memorandum No. 23, Liability Insurance, 15 May 1942; OCD, CAPNHQ, Henry A. Hawgood to All Unit Commanders, memorandum, subject: General Memorandum No. 24, Crash Insurance, 15 May 1942; OCD, CAPNHQ, Henry A. Hawgood to All Unit Commanders, memorandum, subject: General Memorandum No. 25, Accident Insurance, 15 May 1942, Binder "Civil Air Patrol–Establishment of, Charts, Staff, General Memoranda, Training Memoranda," Box 2, Entry 54, CAP, Processed Documents Issued by the OCD, RG171, NARA.

16. OCD, CAPNHQ, Operations Directive No. 13A, Reimbursement Schedules for Coastal Patrol Bases, 1 May 1942, AWS, CAP-NAHC.

17. OCD, CAPNHQ, Aircraft on Active CAP Coastal Patrol Duty by Type of Plane, 28 April 1943, Folder "Submarine-Aircraft Disposition," Box 4, Secretary of War, Office, Expert Consultant to the Secretary of War, Correspondence and Reports, Re: Edward L. Bowles, Anti-Sub Warfare, RG107, NARA.

18. Howard S. Sterne to R. W. Batchelder, memorandum, subject: Distribution of Classified Matter, 27 October 1942, with attachment, "List of Aircraft with Engine Types on Active CAP Coastal Patrol Duty"; and "List of Aircraft with Engine Types on Active CAP Coastal Patrol Duty," 8 October 1942, Reel 38919, AFHRA.

19. "Stinson 10A 'Voyager,'" New England Air Museum, https://www.neam.org/ac-stinson-10a.php; and "Fairchild Model 24-C8F (UC-61J)," National Museum of the United States Air Force, https://www.nationalmuseum.af.mil/Visit/Museum-Exhibits/Fact-Sheets/Display/Article/195812/fairchild-model-24-c8f-uc-61j/.

20. CAPNHQ, Samuel E. Fraim, Manual of Radio Communication Equipment Necessary for a CAP Coastal Patrol, 1 April 1942, BLS, CAP-NAHC.

21. Warner and Grove, *CAP Coastal Patrol Base Twenty-one*, 55–57; *C.C.P. Base 10: Pass in Review!*, 48; Speiser, "'Joe'—Submarine Hunter—1942–1943," 5, CAP-NAHC; McDonald, interview; Swaim, interview; Claude Y. Nanney Jr., CAP Coastal Patrol Base No. 5, Flagler Beach, Florida, May 19, 1942 . . . August 31, 1943 (n.p.: Florida?, 1943), 11–12; Isaac W. Burnham II, "History of CAP Coastal Patrol No. 4," (ca. 1943), 6–8, Reel 44592, AFHRA; Blazich, "North Carolina's Flying Volunteers," 416; Keefer, *From Maine to Mexico*, 170, 214, 283, 375.

22. OCD, CAPNHQ, Harry H. Blee to All Task Force Commanders, memorandum, subject: Radio Stations on Coastal Patrol Task Force Operations, 30 May 1942; OCD, CAPNHQ, "Radio Stations Controlling Coastal Patrol Task Force Operations," 30 May 1942; and OCD, CAPNHQ, Harry H. Blee to All Regional, Wing, and Base Commanders, memorandum, subject: Confidential Letter of Instruction No. 2A Organization of Coastal Patrol Bases, 28 May 1942, Reel 38907, AFHRA.

23. Telegram from Earle L. Johnson to Commanding Officers, CAP, Atlantic City Airport, Morrison Field, and Rehoboth Airport, 7 April 1942, Reel 38907, AFHRA;

and OCD, CAPNHQ, Operations Directive No. 23, Task Forces on Coastal Patrol Duty, 22 June 1942, AWS, CAP-NAHC.

24. Bettie Thompson to Harry H. Blee, 24 April 1942, Reel 38919, AFHRA.

25. Harry H. Blee to Bettie Thompson, 29 April 1942, Reel 38919, AFHRA.

26. Earle L. Johnson to Arlene Davis, 29 May 1942, BLS, CAP-NAHC.

27. Eggenweiler, interview; Keefer, *From Maine to Mexico*, 10.

28. Mary Astor to Kendall K. Hoyt, 28 July 1942, Folder "Civil Air Patrol Historical Documents 1942, Mar–Dec," CAP-NAHC.

29. Eggenweiler, interview; and Compton, interview.

30. Telegram from Cornelius R. Coffey to John F. Curry, 17 December 1941, BLS, CAP-NAHC.

31. Telegram from John F. Curry to Cornelius R. Coffey, 18 December 1941, BLS, CAP-NAHC.

32. "Civil Air Patrol Does the 'Impossible' in Illinois," *Chicago Defender*, 7 March 1942, 1; "Interracial Civil Air Patrol Squadron Set Up in Illinois," *Chicago Defender*, 7 March 1942, 12; "CAP Flyers Sworn in for Duty," *Chicago Defender*, 21 March 1942, 7. The 111th Flight Squadron would be renumbered as Squadron 613-6 by summer 1942.

33. Blazich, "Unsung Civil Rights Milestone," 46–49.

34. Nanney, *CAP Coastal Patrol Base No. 5*, 13, 17.

35. "Aviators End Cross Country Hop," *Chicago Defender*, 19 September 1942, 7; and "Chicago Pilots Complete 2,000-Mile Circuit," *Baltimore Afro-American*, 19 September 1942, 23.

36. Regarding race-related incidents at CAP coastal patrol bases, see Keefer, *From Maine to Mexico*, 89, 243–44.

37. The yearbook for Coastal Patrol Base No. 10, Beaumont, TX, includes several photographs of the African-American base personnel. *C.C.P. Base 10: Pass in Review!*, 43–44, 66.

38. Burnham, "History of CAP Coastal Patrol No. 4," 8, Reel 44592, AFHRA.

39. Almost every coastal patrol base drafted manuals or instructions to help orient and train new personnel. Burnham, "History of CAP Coastal Patrol No. 4," 8, Reel 44592; 12th Task Force, Civil Air Patrol, Brownsville, Texas, memorandum No. 12-01, 17 July 1942; Jack R. Moore, General Orders No. 12, "C.A.P. Coastal Patrol No. 8, Charleston, S.C.," 20 March 1943; Civil Air Patrol Coastal Patrol No. 15, Operations Directive No. 1A to all officers, CCP No. 15, and all pilots, observers, radio operators, and others concerned, 15 October 1942; CAP Coastal Patrol No. 18, Falmouth Airport, Hatchville, Mass., Directive–Memorandum No. 12-01, 7 September 1942; and Civil Air Patrol, Civil Coastal Patrol No. 15, Instructions–Bulletin #1, 4 August 1942, Reel 38920, AFHRA.

40. Bridges, interview; Myers, interview; Arn, interview; and OCD, CAPNHQ, Harry H. Blee, to All Regional, Wing, and Base Commanders, memorandum, subject: Confidential Letter of Instruction No. 5, Organization and Operation of Coastal Patrol Bases, 11 May 1942, Reel 38907, AFHRA.

41. Keefer, *From Maine to Mexico*, 10, 38; and Keil, "Civil Air Patrol Coastal Patrol No. Three."

42. US Air Force, Antisubmarine Command, 215–16; Headquarters I Air Support Command, E. C. Stephan, "Notes for Aircraft on Anti-Submarine Patrol," 15 April 1942; Headquarters Army Air Forces Anti-Submarine Command, Howard Moore, "Whales, Not Submarines," 27 October 1942, AWS, CAP-NAHC; and Timothy Warnock, "The Battle Against the U-Boat in the American Theater" (Washington, DC: Air Force History Support Office, Bolling AFB, 1994), 9–10.

43. Copies of photographs taken by the Navy on 22 June 1942 of an S-class submarine crash-diving were provided to CAP coastal patrol bases to help aircrews identify the appearance of a crash-diving submarine. Some of these images have been mislabeled as CAP photographs of suspected German U-boats. Navy Chief of the Bureau of Aeronautics to Chief of Naval Operations (Training Section), memorandum, subject: Photographs, forwarding of, 7 July 1942, attached to envelope of photographs, Folder "A10(4) A/S Publications (S/M)," Box 1, Tenth Fleet, Anti-Submarine Measures Division, Administrative Files, A10(2)–A10(8), RG38, NARA; and Headquarters I Bomber Command, Howard Moore, memorandum, subject: Information Bulletin No. 16, Series of Pictures Showing an S-Type Submarine Crash Diving, with Surface Speed at Time of Dive–10 Knots, 10 August 1942, Reel A4085, AFHRA.

44. OCD, CAPNHQ, Coastal Patrol Circular No. 50, Instruction in Technique of Antisubmarine Warfare, 24 May 1943; Headquarters Army Air Forces, CAPNHQ, Coastal Patrol Circular No. 53, Instruction in Bombing, 4 June 1943, Reel A4064, AFHRA; Keil, "Civil Air Patrol Coastal Patrol No. Three"; and Myers, interview.

45. Arn, interview; Mosley, *Brave Coward Zack*; McDonald, interview; and Thompson and Thompson, *Palm Beach*, 112.

46. War Diary, Eastern Sea Frontier, April 1942, chap. 1, 1, NARA (via Fold3).

47. King requested Andrews begin planning for a coastal convoy system on 3 April 1942, but on 20 March, King ordered Adm Royal E. Ingersoll, commander-in-chief, US Atlantic Fleet, to initiate coastal convoy routes "at once." War Diary, Eastern Sea Frontier, April 1942, chap. 3, 1, NARA (via Fold3); and Blair, *U-Boat War*, 523.

48. Adolphus Andrews to Ernest K. King, memorandum, subject: Submarine Activities on the Atlantic Coast, 10 April 1942; and Ernest J. King to Adolphus Andrews, memorandum, subject: Submarine Activities on the Atlantic Coast, 14 April 1942, file "A16-3(2) Atlantic," Box 259, Headquarters COMINCH, 1942–Secret, A16-3(1) to A16-3(3), RG38, NARA.

49. War Diary, Eastern Sea Frontier, April 1942, chap. 6, 1–4; May 1942, chap. 7, 1–6, NARA (via Fold3); Morison, *Battle of the Atlantic,* 254–55; Gannon, *Operation Drumbeat,* 386–88; Blair, *U-Boat War,* 525; and Hickam, *Torpedo Junction,* 161–64.

50. Robert Taylor III to Ivan H. Hinderacker, memorandum, 8 April 1942, with attachments "Coastal Patrol Bases–East Coast," and "Report on Civil Air Patrol Missions," Folder "OCD Civil Air Patrol," Box 10, Entry 107A, Budgetary Administration Records for Emergency and War Agencies and Defense Activities, 1939–1949 (file 31.19), RG51, NARA; Robert Taylor III to Lt Behren, 14 April 1942; and Robert Taylor III to Robert M. Webster, 22 April 1942, Reel 38918, AFHRA.

51. James A. Ulio to James Landis, 19 March 1942, Folder "Civil Air Patrol Historical Documents 1942, Mar–Dec," CAP-NAHC.

52. OCD, press release for Saturday morning papers, 28 March 1942; John F. Curry to Bertrand Rhine, 24 March 1942, BLS, CAP-NAHC; and Blazich, "Earle L. Johnson: CAP's Wartime Commander," 9–15.

53. Kenneth N. Gilpin, "I.W. Burnham II, a Baron of Wall Street, Is Dead at 93," *New York Times,* 29 June 2002, section C, 18; and Burnham, "History of CAP Coastal Patrol No. 4" (ca. 1943), 1–4, Reel 44592, AFHRA.

54. Notes of interview of William D. Mason by Robert Neprud, circa 1946–47, Reel 38910; Burnham, "History of CAP Coastal Patrol No. 4" (ca. 1943), 3–4, Reel 44592, AFHRA; Earle L. Johnson to John F. Curry, 15 April 1942, Folder 2, Box 2, ELJ, WRHS; Earle L. Johnson to the Tanker Protection Fund Committee, memorandum, subject: Completion of the CAP Coastal Patrol, 31 January 1944, Reel 38920; notes of interview of Wynant C. Farr by Robert Neprud, circa 1946–47, Reel 38910,

AFHRA; Neprud, *Flying Minute Men*, 14; Mellor, *Sank Same*, 70–74; and War Department, Headquarters of the Army Air Forces, Dudley M. Outcalt to Air Inspector, memorandum, subject: Survey of the Civil Air Patrol, 8 March 1944, 21, Folder 3, Box 5, ELJ, WRHS.

55. Burnham, "History of CAP Coastal Patrol No. 4" (ca. 1943), 5–12, Reel 44592, AFHRA; OCD, CAPNHQ, Operations Orders No. 1, Activation of CAP Coastal Patrols, 30 November 1942, Folder 2, Box 6, ELJ, WRHC; Schlegel, *Virginia on Guard*, 202; and Keefer, *From Maine to Mexico*, 88.

56. CAPNHQ, "Information and Procedure Necessary for the Establishment of a Base for Coast Patrol Missions," undated (believed late June–July 1942), BLS, CAP-NAHC.

57. CAPNHQ, "Information and Procedure Necessary for the Establishment of a Base for Coast Patrol Missions," undated (believed late June–July 1942), BLS, CAP-NAHC.

58. OCD, CAPNHQ, Harry H. Blee to All Regional, Wing, and Base Commanders, memorandum, subject: Confidential Letter of Instruction No. 1, Operation of Coastal Patrol Bases, 14 April 1942, Reel 38907, AFHRA.

59. OCD, CAPNHQ, Harry H. Blee to All Regional, Wing, and Base Commanders, memorandum, subject: Confidential Letter of Instruction No. 2, Organization of Coastal Patrol Bases, 16 April 1942, Reel 38907, AFHRA.

60. OCD, CAPNHQ, Operations Directive No. 23, Task Forces on Coastal Patrol Duty, 22 June 1942, AWS, CAP-NAHC.

61. Craven and Cate, *Plans and Early Operations*, 527–31; and Blair, *U-Boat War*, 515–16.

62. Blair, *U-Boat War*, 535, 544, 568; and Doenitz, *Memoirs*, 218–19.

63. Robert A. Lovett to Henry H. Arnold, memorandum, 18 April 1942; Henry H. Arnold to Follett Bradley, memorandum, subject: Equipment of the C.A.P. Squadron at Morrison Field, 20 April 1942; and Follett Bradley to Henry H. Arnold, memorandum, 28 April 1942, Folder "Military Official Nos. 270-273," Box 177, HHA, LOC.

64. Follett Bradley to Henry H. Arnold, memorandum, 28 April 1942, Folder "Military Official Nos. 270-273," Box 177, HHA, LOC.

65. "Civilian Pilots Aid in Sinking U-boats: Air Force Reveals Spotting of Enemy Raiders Off Coast by Amateur Fliers," *New York Times*, 20 April 1942, 23. Days later CAP National Headquarters distributed the announcement by the general to its entire membership. OCD, *CAP Bulletin* 1, no. 13, 24 April 1942, Folder 6, Box 6, ELJ, WRHS.

66. Cressman, *Official Chronology*, 88.

67. Public Relations Officer, Air Forces, Eastern Defense Command and First Air Force, "For Release Monday, April 20, 1942," Folder "Civil Air Patrol Historical Documents 1942, Mar–Dec," CAP-NAHC.

68. "Civil Air Patrol: America's Private Pilots Are Mobilized for War," *Life* 12, no. 17 (27 April 1942): 63–66.

69. OCD, CAPNHQ, Harry H. Blee, Operations Directive No. 6, "Table of Organization for Civil Air Patrol," 10 March 1942, AWS, CAP-NAHC; and OCD, CAPNHQ, Special Bulletin, 1 May 1942, Folder "Civil Air Patrol 15 July 1942," Entry 16A, General Correspondence, 1941–May 1945, Civil Air Patrol, RG171, NARA. The reason for the lack of photography was to guard against the possibility of photographs disclosing the movements of friendly surface vessels. Harry H. Blee to Louis E. Boutwell, memorandum, subject: Cameras on CAP Coastal Patrol Bases, 14 October 1942, Folder "AAF ASC Policy," CAP-NAHC.

70. OCD, *CAP Bulletin* 1, no. 3, 13 February 1942, Folder 6, Box 6, ELJ, WRHS.

71. OCD, CAP, Special Bulletin to Wing Commanders for transmittal to all CAP intelligence officers, memorandum, subject: Intelligence and Public Relations Procedure, 30 March 1942; and OCD, *CAP Bulletin* 1, no. 13, 24 April 1942, Folder 6, Box

6, ELJ, WRHS; and OCD, CAPNHQ, Earle L. Johnson to Wing and Group Commanders, memorandum, subject: Intelligence Officers (GM-10), 13 February 1942, Binder "Civil Air Patrol–Establishment of, Charts, Staff, General Memoranda, Training Memoranda," Box 2, Entry 54, CAP, Processed Documents Issued by the OCD, RG171, NARA.

72. Kendall K. Hoyt to Earle L. Johnson, memorandum, subject: Protection of Secrecy of CAP Coastal Patrol, 28 October 1942, Reel 38919, AFHRA; Office of Civilian Defense, Civil Air Patrol National Headquarters, Kendall K. Hoyt to all CAP coastal patrol and liaison patrol commanders, special memorandum, subject: Publicity on Accidents, 23 March 1943, Coastal Patrol Base No. 16 Collection (CP16), CAP-NAHC; and OCD, CAPNHQ, Harry H. Blee, Coastal Patrol Circular no. 40, "Safeguarding Intelligence and Communications Offices, 7 April 1943, Binder "Civil Air Patrol," Box 1, Entry 54, Processed Documents Issued Serially by the Civil Air Patrol, RG171, NARA.

73. By July 1943, the CAP coastal patrol bases cost an average of $29,657 to operate. Louis F. Licht Jr. to Henry H. Arnold, memorandum, subject: Cost of CAP Coastal Patrol, July 1943, 8 October 1943, Reel 38920, AFHRA.

74. Harry H. Blee, OCD, CAPNHQ, memorandum, subject: Memorandum on Civil Air Patrol Operations, 13 May 1942, BLS, CAP-NAHC; House Committee on Appropriations, *Supplemental National Defense Appropriation Bill for 1943: Hearings on H.R. 7319*, Part 1, 77th Cong., 2nd sess., 1942, 824–25. Some of these funds were also intended for the Fourth Task Force and approval for the other bases arrived around 12 May. Harry H. Blee, OCD, CAPNHQ, memorandum, subject: Memorandum on Civil Air Patrol Operations, 13 May 1942, BLS, CAP-NAHC; and CAPNHQ, Operations Orders No. 1, "Activation of CAP Coastal Patrols," 30 November 1942, Folder 2, Box 6, ELJ, WRHC; Nanney, *CAP Coastal Patrol Base No. 5*, 10.

75. Senate Committee on Appropriations, *Supplemental National Defense Appropriation Bill for 1943: Hearings on H.R. 7319*, 77th Cong., 2nd sess., 1942, 101–4, 253; House Committee on Appropriations, *Supplemental National Defense Appropriation Bill for 1943: Hearings on H.R. 7319*, Part 1, 77th Cong., 2nd sess., 1942, 800–802; John Fisher, "Arnold Faces Quiz on Limiting *Civil Air Patrol*: Increasing Sub Toll Stirs Senators to Actions," *Chicago Daily Tribune*, 9 July 1942, 3; and Senate Committee on Appropriations, *Military Establishment Appropriation Bill, 1943*, 77th Cong., 2nd sess., 1942, S. Rep. 1527, 33–34.

76. Laurence S. Kuter to Budget Office, memorandum, subject: Air Force Utilization of Civil Air Patrol, 4 July 1942, Reel 38918, AFHRA.

77. James A. Ulio to Brehon B. Somervell, memorandum, subject: Organization of Elements of Civil Air Patrol Under Military Jurisdiction, 29 April 1942; and James A. Ulio to Henry H. Arnold, memorandum, subject: Organization of Elements of Civil Air Patrol under Military Jurisdiction, 29 April 1942, Folder "SAS 324.3–Civil Air Patrol," Box 93, HHA, LOC.

78. War Department, *Army Specialist Corps Regulations (Tentative) 1942*, 2.

79. Millard F. Harmon to Henry H. Arnold, memorandum, subject: Organization of Elements of Civil Air Patrol under Military Jurisdiction, 1st Indorsement, 5 May 1942, Folder "SAS 324.3–Civil Air Patrol," Box 93, HHA, LOC; and Randolph Williams to Follett Bradley, memorandum, subject: 1st Indorsement, 25 July 1942, Reel A4064, AFHRA.

80. Portion of a memorandum from Harry H. Blee to unknown, 2 May 1942, Binder "Legal Status, Administrative Concepts, and Relationships of the Civil Air Patrol, 1941 to 1949," CAP-NAHC.

81. Lawrence J. Kuter to the Budget Officer, War Department, memorandum, subject: Air Force Utilization of Civil Air Patrol, 4 July 1942, Reel 38918, AFHRA.

82. Gannon, *Operation Drumbeat*, 387; Blair, *U-Boat War*, 568–69, 573–74; and War Diary, Eastern Sea Frontier, May 1942, chap. 1, 1–4, chap. 4, 1–10, NARA (via Fold3).

83. OCD, CAPNHQ, Harry H. Blee to All Regional, Wing, and Base Commanders, memorandum, subject: Confidential Letter of Instruction No. 5, Organization and Operation of Coastal Patrol Bases, 11 May 1942; and OCD, CAPNHQ, Harry H. Blee to all Regional, Wing, and Base Commanders, memorandum, subject: Confidential Letter of Instruction No. 9, Coastal Patrol and Convoy Operations, 23 May 1942, Reel 38907, AFHRA.

84. Doenitz, *Memoirs*, 220–21; Roskill, *Period of Balance*, 102; Murray and Millett, *A War to Be Won*, 251–52; Craven and Cate, *Plans and Early Operations*, 530–31; Morison, *Battle of the Atlantic*, 137–42; Blair, *U-Boat War*, 578–88; and US Air Force, *Antisubmarine Command*, 35.

85. Keil, "Civil Air Patrol Coastal Patrol No. Three."

86. Keil, "Civil Air Patrol Coastal Patrol No. Three"; Mellor, *Sank Same*, 101; and KTB for Fifth War Patrol of *U-564*, entry for 4 May 1942, http://uboatarchive.net/U-564/KTB564-5.htm. 1st Lt Jake E. Boyd, 1st Lt Carl N. Dahlberg, Capt Ted F. Keys, 1st Lt Wallace R. King, and Maj Wright Vermilya Jr. received commendations.

87. Keil, "Civil Air Patrol Coastal Patrol No. Three"; Blair, *U-Boat War*, 546; and Cremer, *U-Boat Commander's Handbook*, 70–74.

88. Keil, "Civil Air Patrol Coastal Patrol No. Three"; and Mellor, *Sank Same*, 102.

89. Keil, "Civil Air Patrol Coastal Patrol No. Three"; Neprud, *Flying Minute Men*, 18; Mellor, *Sank Same*, 98–101; and Thompson and Thompson, *Palm Beach*, 110–11.

90. KTB for Fifth War Patrol of *U-109*, entry for 6 May 1942, http://uboatarchive.net/U-109/KTB109-5.htm. *U-109*'s KTB mentions maneuvering in water 16–18 meters deep and that the targeted tanker was aground in 8 meters of water. The submarine fired a torpedo, which missed.

91. Hickam, *Torpedo Junction*, 209–17; Blair, *U-Boat War*, 538, 546, 568; and Doenitz, *Memoirs*, 220.

92. Keil, "Civil Air Patrol Coastal Patrol No. Three"; Blair, *U-Boat War*, 568; Wiggins, *Torpedoes in the Gulf*, 60; Morison, *Battle of the Atlantic*, 139; and "Neighbor Republic Acts After Two Vessels Are Sunk by Submarines," *Washington Post*, 23 May 1942, 1–2.

93. Ralph A. Bard to Frederick J. Horne, 9 May 1942, Box 165, HHA, LOC.

94. Henry H. Arnold to George C. Marshall, memorandum, subject: Submarines Off Florida Coast, 9 May 1942, Box 165, HHA, LOC. The report Arnold received may have originated from a direct telephone call between Vermilya at the Third Task Force and Arnold. Mosley, *Brave Coward Zack*, 55; and Thompson and Thompson, *Palm Beach*, 111.

95. Henry H. Arnold to George C. Marshall, memorandum, subject: Anti-Submarine Operations, 10 May 1942, Box 165, HHA, LOC.

96. Reed G. Landis to John F. Curry, memorandum, 13 February 1942, BLS, CAP-NAHC.

97. C. I. Stanton to Donald H. Connolly, memorandum, subject: Carriage of explosives in civil aircraft, 14 May 1942, Folder "020 War Department/Vol. 7 [Folder 1 of 2—February–May 1942]," Box 6, Entry UD1, Civil Aeronautics Administration, RG237, NARA.

98. Telegram from Henry H. Arnold to Follett Bradley, undated (estimated 10 May 1942), Folder "Military Official Nos. 270–273," Box 177, HHA, LOC. The origi-

nal document, despite reading "confirming telephone instructions this date," has no listed date.

99. Card index, "Puddle Jumpers," 11 May 1942, Box 57, HHA, LOC.

100. Headquarters I Ground Air Support Command to Civil Air Patrol Authorities, "Letter of Instructions Number 4," 16 May 1942 (for Fifth, Sixth, Seventh, Eight Task Forces), Reel 38920, AFHRA. Emphasis added. The new task forces were located at Daytona/Flagler Beach, Florida (5th); St. Simons Island, Georgia (6th); Miami, Florida (7th); and Charleston, South Carolina (8th). On 20 May, I Ground Air Support Command issued a new letter of instructions to the First and Second Task Forces with a mission statement that included "to take action to destroy, by bombing attacks, any enemy submarines sighted." The instructions noted that no CAP unit would carry or drop live bombs or depth charges until specifically directed by the Army. Headquarters, I Ground Air Support Command, John P. Doyle Jr. to Civil Air Patrol Authorities, First Task Force, Atlantic City, New Jersey, Second Task Force, Rehoboth, Delaware, Letter of Instructions Number 1, 20 May 1942, Reel A4064, AFHRA. The War Department redesignated I Air Support Command as I Ground Air Support Command in April 1942. Maurer, *Air Force Combat Units of World War II*, 440.

101. Dwight D. Eisenhower to Henry H. Arnold, memorandum, subject: Action to Improve Anti-Submarine Operations, 20 May 1942, Reel A4063, AFHRA.

102. Headquarters I Ground Air Support Command to Civil Air Patrol Authorities, 9th Task Force, New Orleans, La., "Letter of Instructions Number 8)," 29 June 1944, Reel 38920, AFHRA.

103. OCD, CAPNHQ, Henry A. Hawgood to All Unit Commanders, memorandum, subject: General Memorandum No. 43, Extension of Liability Insurance, Crash Insurance and Accident Insurance Policies to Cover Losses Incurred while Aircraft are Equipped with Bomb Racks, 13 July 1942, Binder "Civil Air Patrol–Establishment of Charts, Staff, General Memoranda, Training Memoranda," Box 2, Entry 205, Processed Documents Issued by the OCD–CAP, RG171, NARA; and OCD, CAPNHQ, Operations Directive No. 13B, Reimbursement Schedules for Coastal Patrol Missions, 14 July 1942, AWS, CAP-NAHC.

104. E. H. Eddy, Headquarters I Ground Air Support Command, to Commanding General, Air Forces, E.D.C. and First Air Force, memorandum, subject: Ordnance Service for Civil Air Patrol Units, 26 June 1942; William P. Winslade to Henry H. Arnold, memorandum, subject: Ordnance Service for Civil Air Patrol Units, 1st Indorsement, 30 June 1942; F. H. Monahan to Commanding General, Air Forces, E.D.C. and First Air Force, memorandum, subject: Ordnance Service to Civil Coastal Patrol Units, 11 August 1942; Charles L. Carlson to Commanding General, I Bomber Command, memorandum, subject: Ordnance Service to Civil Coastal Patrol Units, 3rd Indorsement, 3 September 1942; and Harry H. Blee to Louis E. Boutwell, memorandum, subject: Ordnance Supplies and Services for Civil Air Patrol Coastal Patrols, 4 November 1942, Reel A4064, AFHRA.

105. OCD, CAPNHQ, Operations Directive No. 23, "Task Forces on Coastal Patrol Duty," 22 June 1942, AWS, CAP-NAHC.

106. OCD, CAPNHQ, Harry H. Blee to All Regional, Wing, and Base Commanders, memorandum, subject: Confidential Letter of Instruction No. 2A, Organization of Task Forces on Coastal Patrol Duty, 28 May 1942, Reel 38907, AFHRA; OCD, CAPNHQ, Operations Directive No. 13B, Reimbursement Schedules for Coastal Patrol Missions, 14 July 1943, AWS, CAP-NAHC; Keefer, *From Maine to Mexico*, 173–74, 247–49; and Herman Kehrli to Bernard L. Gladieux, memorandum, subject: Civil Air Patrol Relationships, 26 May 1942, Folder "OCD Civil Air Patrol,"

Box 10, Entry 107A, Budgetary Administration Records for Emergency and War Agencies and Defense Activities, 1939–1949 (file 31.19), RG51, NARA. Photographic records from some of the coastal patrol bases indicate guards also carried an array of revolvers and semiautomatic pistols. In several instances, Army or National Guard personnel issued CAP guards standard-issue .45-caliber automatic pistols, .30-06-caliber bolt-action rifles, and Thompson submachine guns.

    107. OCD, CAPNHQ, Operations Directive No. 23, "Task Forces on Coastal Patrol Duty," 22 June 1942; and OCD, CAPNHQ, Operations Directive No. 23A, "CAP Coastal Patrols," 26 August 1942, AWS, CAP-NAHC.

Chapter 5

From Maine to Mexico

I definitely think that the Civilian Air Patrol has immeasurably helped on eliminating the submarine menace, and further express a desire to see these planes flying overhead the next time I go to sea.

—Charles R. Berg, master, MS *Southern Sun*, 13 November 1942

As the U-boat offensive shifted into the Gulf of Mexico, CAP followed the Army and Navy's forces. As defenses and the introduction of convoys along the East Coast limited target opportunities for Germany's gray wolves, the hunters moved south. "The Americans, apparently, had not anticipated the appearance of U-boats in such far distant parts of the Caribbean as the Gulf of Mexico," recalled Doenitz. "Once again we had struck them in 'a soft spot.'"[1] The U-boat command swiftly transferred six boats from the East Coast to southern waters and routed four additional boats to the Caribbean. In May, U-boat attacks rose sharply and losses in the Gulf Sea Frontier doubled those of the Eastern Sea Frontier. Through the support of specialized Type XIV tanker/supply boats, dubbed *milchkühe* (milk cow), Type VII and IX U-boats could enjoy longer patrols in American and Caribbean waters through replenishment of fuel oil, food, torpedoes, and spare parts.[2]

Within the Gulf, oil tankers full of the modern war's lifeblood steamed unescorted from ports in Texas and Louisiana. Much like the situation Andrews faced in January, Capt Russell S. Crenshaw, commander, Gulf Sea Frontier, found himself with a pittance of resources when he took command on 6 February. His initial defenses consisted of only one converted yacht, three Coast Guard cutters, and 35 Army and Coast Guard aircraft, over half unarmed. U-boat operations in the Gulf of Mexico in May sank 41 vessels—with tankers representing 55 percent of the total—almost double the April losses in the Eastern Sea Frontier (23 ships). The combined sea frontier losses of 46 ships in May 1942 represented the worst month in the war; for the Gulf Sea Frontier, these were the deadliest American waters in the entire war. Over the previous three months, Crenshaw had barely managed to increase his forces with the addition of two aged destroyers, two more converted yachts, six Coast Guard cutters, a naval air detachment, and a smattering of Army aircraft.[3] The additions did little to thwart the enemy and consequently brought a change in leadership. On 3 June, King appointed Rear Adm James L. Kauffman as Gulf Sea Frontier commander, replacing Crenshaw the following day. With experience hunting U-boats in the North Atlantic, Kauffman bore heavy expectations to produce results and staunch the flow of blood and oil.[4]

Before his relief, Crenshaw had requested reinforcements from the Army Air Forces as losses to U-boats mounted in May. On 26 May, Bradley established the Gulf Task Force under I Bomber Command, composed of 20 B-18 bombers, the 66th and 97th Observation Squadrons, and CAP's Third, Fifth, and Seventh Task Forces.[5] In late June, I Ground Air Support Command had proposed establishing nine more CAP coastal patrol bases in Maine, New York, Mississippi, Louisiana, and Texas. These bases and their estimated 155 aircraft, rigged with bomb racks and radios, would allow the Army to release 12 observation squadrons from coastal patrol duty for joint training with the Army Ground Forces to prepare for overseas assignment.[6] By 25 June CAP National Headquarters had activated task forces at Grand Isle, Louisiana; Beaumont, Texas; and Pascagoula, Mississippi. In early July, the Army attached the 128th and 124th Observation Squadrons to the Gulf Task Force to help stand up and train these new CAP coastal patrol units. By 7 July, all three task forces had begun patrols providing aerial coverage of merchant traffic moving along the gulf rim.[7]

The Ninth Task Force commenced patrol operations on 6 July. Flying out of New Orleans' Lakefront Airport, CAP aircraft covered shipping lanes from approximately Port Eads to the western end of Marsh Island. I Ground Air Support Command prioritized patrols of "maximum density" during daylight hours over the entrance to the southwest channel of the Mississippi River and the steamer lanes west- and eastward from the entrance to the river.[8] With traffic heavy at the city field, the men and women of the "Fighting Nine" relocated 50 miles away to Grand Isle. Readying this base, one-time home to famed French pirate Jean Lafitte, made the challenges encountered at Parksley seem paltry. The two runways were shell-covered roads, one less than 50 yards from the ocean. The dilapidated, rat-infested Oleander Hotel provided barracks and served as an administration building. Water came from a cistern with a silk stocking tied around faucets to skim off the mosquito larvae. With help from locals, base personnel drove three 90-foot poles into the ground to serve as an antenna for the base radio. The state road department leveled and improved the roads, producing two 900- x 40-foot runways. A large canvas sheet draped over a telephone pole served as the base's first hangar, with mechanics working at night by flashlight with oil and driftwood fires for heat. Primitive conditions aside, mere minutes after takeoff, aircraft came upon the war, seeing "what seemed like an endless stream of wreckage, litter and debris" and trails of black oil leading to the sunken remains of tankers.[9]

Figure 16. Aerial view of the Ninth Task Force, Grand Isle, Louisiana. Note the proximity of the runway to the Gulf of Mexico. (Photograph courtesy of the Morse Center.)

CAP National Headquarters sought volunteers and aircraft to meet the Army's request for new coastal patrol bases. For the former, headquarters urged wing commanders to submit paperwork on available equipment and personnel.[10] Individual members were asked "to consider how much time he or she can devote to active duty missions," for one to three months of continuous service per year.[11] To obtain aircraft, CAP opted to appeal to members' patriotism: "Airworthy equipment is

Figure 17. Tech Sgt Addis H. McDonald of North Little Rock, Arkansas, holds an oil-soaked life jacket found washed ashore on Grand Isle. (Photograph courtesy of the Morse Center.)

wanted and if you have it you will serve your country by making it available."[12] Under Johnson's signature, hundreds of aircraft owners across the country received letters, which opened with, "As you no doubt know, we are performing coastal patrol operations. This duty is very essential and important in relieving the Armed Forces, and as an owner of an aircraft with over 200 horsepower, we hope you would be interested in letting us use your aircraft and if possible, yourself, for a period of not less than 30 days for this duty." Assuredly, if the owner could not fly the aircraft, CAP had plenty of competent pilots ready to ferry the aircraft to a task force in need. "I do not know what finer contribution you could make as a member of the Civil Air Patrol to the war effort than to donate your aircraft," continued Johnson, "and, of course, yourself if possible, to this service."[13] Within months CAP would further refine its approach to document its registered aerial fleet and search records to help locate new aircraft to fill the ever-critical need for qualified airframes.[14]

The new slaughter of tankers in the Gulf of Mexico and the Caribbean brought long-simmering doctrinal conflicts about antisubmarine warfare between the Army and Navy to a boil. With the outcome of the fighting in North Africa and the Soviet Union still in doubt for the Allies, the antisubmarine situation brought forth an exchange of heated memoranda between Marshall and King. The principal issue of contention was jurisdiction for coastal defense operations. Discussion of the differing supporting elements regarding availability and employment of antisubmarine assets, however, exposed critical philosophical operational differences between the Army and the Navy regarding the use of land-based aircraft.[15] Marshall wrote King and candidly stated that "the losses by submarines off our Atlantic seaboard and in the Caribbean now threaten our entire war effort." After listing several loss statistics, the general pointedly asked: "has every conceivable improvised means been brought to bear on this situation?"[16] Marshall feared that another month or two of losses to U-boats would cripple the nation's means of transportation, thwarting the ability to move sufficient forces to bear against the enemy in critical theaters of the war.

Known to some for his bluntness and sharp temper, King parried Marshall's message with a rather measured reply. He acknowledged that "if we are to avoid disaster not only the Navy itself but also all other agencies concerned must continue to intensify the antisubmarine effort." The admiral outlined the Navy's actions and those of the Army since the commencement of hostilities, notably the "extemporization by taking on the civil aviation patrol." King deemed the overall situation "not hopeless." He continued:

> We know that a reasonable degree of security can be obtained by suitable escort and air coverage. . . . But if all shipping can be brought under escort and air cover our losses will be reduced to an acceptable figure. I might say in this connection that escort is not just *one* way of handling the submarine menace; it is the *only* way that gives any promise of success. The so-called patrol and hunting operations have time and again proved futile. We have adopted the "Killer" system whereby contact with a submarine is followed up continuously and relentlessly—this requires suitable vessels and planes which we do not have in sufficient numbers.[17] (emphasis in original)

From the perspective of doctrine, the Navy's "defensive" approach to bringing the submarine to the convoy contrasted with the Army's "offensive" operation of a single centralized command controlling mobile units, which could shift from location to location as required. Arduous, seemingly unproductive patrols would be replaced by hunter-killer forces of Army bombers able to find, fix, and destroy a submarine needle in the ocean haystack.

CAP's coastal patrol operations, although trained by the Army, conformed more with the Navy's doctrine. Flying out of fixed bases, CAP aircraft flew the dull routine observation deterrence patrols rather than engaging in hunter-killer actions. The light civilian aircraft conducted escorts for convoys or individual merchant ships at the Navy's request. The Army's decision to arm CAP provided a potential means to offensively strike back should opportunity allow, but otherwise CAP provided what Andrews and Kauffman both required: aerial deterrence. With defended convoys operational in the Gulf by August, Doenitz again withdrew his forces to more profitable waters.[18]

With the waters off the East and Gulf Coasts at least devoid of prowling U-boats, the Navy tacitly acknowledged CAP's coastal patrol service. Now–Vice Admiral Andrews noted that while convoying played the primary part in reducing losses, "support and offensive action by aircraft played an important role."[19] The Bureau of the Budget casually remarked that month how "it appears that the Army and Navy have wholeheartedly endorsed the work of the Civil Air Patrol."[20] Adm King kept his own counsel regarding CAP, but his silence acknowledged that the civilian volunteers had proved up the task at hand.

While CAP found itself serving two masters, the Army pulled CAP further into its organizational culture—through uniform appearance. Before CAP's establishment, Wilson wanted something that was attractive while also practical and affordable to civilians, first suggesting a dark blue uniform coat with gray slacks or skirt and a gray shirt with black tie and shoes.[21] Wilson's assistants, Helen Rough and Cecile Hamilton, consulted off-the-rack options at Sears Roebuck, JC Penney, and Montgomery Ward and recommended Army-style twill in shades of blue, teal green, and smoke khaki.[22] Wilson proposed a khaki uniform as it represented the most serviceable of all colors which was easily procured or already owned by potential volunteers, and would "convey the sense of national defense significance."[23]

LaGuardia, however, wanted members to wear a distinctive uniform rather than a copy of an existing military uniform, to remain distinctly OCD in shades of blue or grey-blue with black accessories.[24] The OCD director also sought influence on the design of the CAP pilot wings, personally requesting the design resemble those worn by the pilots of Italy's *Regia Aeronautica*.[25] On 3 December 1941, Blee met with members of the Quartermaster General's office to finalized details of CAP's uniform. The parties agreed upon a distinctive garrison cap in blue matching the colors of worsted trousers for men (and culottes for women) and a broadcloth shirt. A darker blue, worsted, single-breasted coat would feature no shoulder straps but have a false belt similar to that of Army Air Forces officer's uniform. A plain black tie and shoes would complete the ensemble.[26]

When the war commenced, CAP's uniform color changed out of logistical necessity. The uniform color scheme changed from blue to brown as flying cadet uniforms required the same dye and cloth.[27] The Quartermaster General–approved CAP uniform of February 1942 for commissioned male and female personnel consisted of a dark brown service coat similar to those worn by Army Air Forces officers, albeit using less fabric with brown plastic buttons; a light brown braid extended from the inner seam to the outer seam of the sleeve at a 45 degree angle from a point three inches above the end of the sleeve. Trousers would be light brown, flat-front commercial pattern without cuffs for men and of the same color in culottes for women. The shirts, distinctive garrison cap, and neckties would also be a light brown commercial pattern, paired with brown waist belts and shoes. The authorized insignia grade consisted of small blue-and-white embroidered tapes worn on the left sleeve a half an inch below the CAP shoulder insignia on the outer garment. These listed titles including "Wing Commander," "Wing Staff," "Group Commander," "Squadron Commander," and so forth.[28] All other noncommissioned CAP personnel wore either matching khaki or olive drab shirts and trousers with a black necktie and tan shoes and belt. The colors evolved between February and April to an entirely commercially-available khaki and tan ensemble.[29]

With the coastal patrol effort succeeding and the War Department increasingly funding CAP operations, civilian volunteer uniforms changed again. Blee had been quietly working on getting CAP into Army fabrics and colors since February. In a message to the Quartermaster General's office, Blee brought up the example of the members of the Virginia Wing, notably those individuals also serving in the Virginia Flying Corps, the aerial component of the Virginia Protective Force, the state militia. Organized in October 1941, the flying corps wore Army-style uniforms approved by the Army's Adjutant General, with "VPF" instead of "US" on their collar insignia and a maroon sleeve braid sewn above the sleeve cuff. After CAP's establishment in December, Maj Allan C. Perkinson, commander of the Virginia Flying Corps, also became commander of CAP's Virginia Wing. In January 1942, CAP and the Virginia Flying Corps reached an agreement whereby the members of the latter would enroll in the former. CAP permitted the flying corps members to wear CAP insignia on their protective force uniforms.[30]

The two-tone brown and khaki uniform options of CAP remained in force as the coastal patrol operations commenced. On 9 June, after various discussions within the War Department and in view of the volunteer auxiliary work "in the field" (that is, the coastal patrol operation) and difficulties obtaining the approved brown uniform, CAP requested wear of the standard Army service uniforms and grade. The uniform would have distinctive red shoulder loops and matching sleeve braid, the CAP shoulder sleeve insignia, silver buttons, and silver CAP lapel insignia. The latter would replace the gold "US" Army Air Forces letters with silver "CAP" letters, and silver wing-and-propeller insignia for the Army's gold and silver equivalent.[31] With the concurrence of Arnold and the Quartermaster General, the War Department approved the request at the month's end.[32]

Wear of Army rank insignia remained the last hurdle for CAP's uniform revision. At Blee's request, the Quartermaster General's office developed an array of unique rank designs and color variations.[33] The distinctive red shoulder loop, however, construed as a "distinctive mark," proved sufficient for CAP to not conflict with Section 125 of the National Defense Act of 1916.[34] In July, CAP National Headquarters announced the wear of the regular Army uniforms with distinctive insignia and the qualifications for rank and grade.[35] As Johnson explained to CAP's rank and file, wear of the uniform "is a privilege granted to no other organization. No one else has been accorded the honor of wearing the Air Corps wing and propeller emblem which, in silver, is prescribed for officers of CAP. No other unit outside the armed services may wear the U.S. which appears on the shoulders of our men." Johnson further noted the ability for members to hold titles of rank and wear the same insignia as worn by Army officers. "This trained corps, unique in the history of warfare, has won good will of the public to a degree which is heartening," he added.[36]

The Army uniform with distinctive insignia combined with identical rank insignia for officers (enlisted wore Army chevrons of embroidered khaki on red fabric) militarized the appearance of CAP personnel. These uniforms represented a far cry from LaGuardia's desire to make them distinctively OCD. Although many CAP members loathed the red shoulder loops, the War Department authorization of the rank grade for the civilians further acknowledged the demonstrated professional discipline of CAP's members, notably the military-style organization of the coastal patrol effort. On the latter, CAP National Headquarters in late 1942 introduced a distinctive active duty emblem for coastal patrol personnel. This black and gold patch of a bomb falling on a submarine conning tower replaced the rather innocuous and mundane "V" patches worn by early volunteers.[37]

Figure 18. Capt Warren E. Moody, engineering officer for Coastal Patrol Base No. 17, Suffolk, Riverhead, New York, wears the CAP officers' winter uniform circa 1943. The two small stripes below the coastal patrol emblem represent 12 months of active duty service. (Photograph courtesy of the Morse Center.)

CAP's new uniforms would soon make appearances along the Gulf Coast. Army funding facilitated CAP activation of five more task forces, four of which provided coverage from the Mexican border to the entirety of Florida.[38] A new base at Manteo, North Carolina, seemed an odd choice considering that on 19 July 1942, Doenitz withdrew the last two U-boats off Cape Hatteras (*U-754* and *U-458*) and transferred operations to the mid-Atlantic.[39] Postwar, Doenitz remarked that despite his shift in priority, "American waters were nevertheless still worthy of exploitation in any area in which the defensive system was found to be still defective."[40]

While the public remained unaware of the withdrawal of Doenitz's U-boats, stories about the CAP coastal patrol began to appear in the nation's newspapers. Nationally, weekly installments of Zack Mosley's popular, nationally syndicated cartoon, *Smilin' Jack*, featured CAP aircraft and characters. Mosley, a native of Hickory, Oklahoma, joined CAP in December 1941 and became a coastal patrol pilot at the Third Task Force on 2 April 1942. In between his patrol flights, he set up his studio at a hangar at the base airport and began to compile material to incorporate coastal patrol and other CAP mission stories into the *Smilin' Jack* storyline. Although initially unable to share military particulars of the coastal patrol experience, Mosley's cartoons introduced Americans to the creativity and daring of the CAP volunteers.[41]

Cartoons also provided the medium for creating unique insignia for coastal patrol bases. Mosley designed a humorous design for the Third Task Force featuring a little yellow aircraft straining under the weight of a massive bomb. His design eventually adorned all the assigned Third Task Force aircraft. Mosely thereafter received requests to design additional CAP and Army Air Force insignia. His design for the Second Task Force at Rehoboth featured a diving Delaware blue hen dropping bombs on a U-boat conning tower. Base No. 17 at Suffolk, Riverhead, New York, used a Mosley-designed insignia consisting of a tired, hitch-hiking aircraft in the clouds.[42] Walt Disney Productions further assisted CAP coastal patrol units, creating colorful, distinctive insignia for Bases No. 14, No. 16, and No. 18.[43]

Outside of the funny pages, accounts of CAP coastal patrol displays of airmanship and rescue operations became national news. Mosley himself narrowly missed making a headline on a dusk patrol flight of 24 April 1942 when his Rearwin Cloudster swallowed a valve 10 miles off Vero Beach, Florida. With his engine dying and smoking, he and his observer, John Prince, just managed to reach the beach at only 300 feet altitude. Skimming at treetop level, the men landed at the nearby Vero Beach Airport, managing to dodge the equipment and workers constructing the field. Mosley's aircraft was the first aircraft on coastal patrol duty to lose an engine in flight, but thankfully he and Prince sustained no injuries and saved the aircraft.[44]

In the afternoon of 21 July, a two-ship flight out of Rehoboth armed with practice bombs commenced an uneventful patrol from Delaware down the Virginia coast. Cruising 15 miles offshore at an altitude of 400 feet, the patrol consisted of two Fairchild 24s armed with practice bombs, the lead ship crewed by 1st Lt Carl L. Virdin, pilot, of Lakewood, Ohio, with 1st Lt Shelley S. Edmondson of Selkirk, New York, as observer and the sister ship piloted by 1st Lt Henry T. Cross with observer 1st Lt Charles E. Shelfus, both of Columbus, Ohio. At 1654 hours while 15 miles off Assateague Island (approximately 27.5 miles northeast of Chincoteague) Virdin noticed Cross and Shelfus dropping out of formation and entering into a spin. Edmondson called in a distress call to base as the stricken Fairchild 24, registration number NC19144, completed one and a half turns before it struck the water vertically at full throttle. The shock of the impact tore the fabric from the skin of the fuselage.[45]

The almost vertical plane began to slowly sink nose downward. Virdin and Edmondson circled above; after a few minutes they saw one of the two crewmembers rise to the surface with an inflated life jacket. He climbed above the horizontal stabilizer and waved. The other crewman never emerged. Five minutes after impact the fuselage completely sank apart from the upper portion of the vertical stabilizer. The lone survivor paddled a short distance from the aircraft and waved with both hands before stretching out in a floating position. While Virdin continued to circle, Edmondson dropped a life preserver, two life jackets, and a smoke marker to spot the survivor's position in the rough water. The marker, however, failed to ignite, and then the men opted to keep the oil and gasoline slick from the crash in constant sight. Virdin flew a short distance off to safely jettison his practice bomb then returned to slowly circle the survivor, rocking his wings and trying to remain visible. Half an hour after impact, the wreckage sank from sight.[46]

Back at the Second Task Force, Edmondson's distress call reached the desk of the commander, Capt Hugh R. Sharp Jr. of Greenville, Delaware. Noticing 1st Lt Edmond I. Edwards of Newark, Delaware, standing nearby listening to the radio, Sharp turned to him and said,

"Eddie, you and I are going out there and see what we can see."[47] Edwards grabbed a nearby coil of rope and some makeshift position markers. Climbing into the base's Sikorsky S-39B, Sharp took the controls on the left while Edwards manned the radio on the right.[48]

At 1848 hours, the Sikorsky arrived at the crash site. Edmondson dropped two more smoke markers to mark the last position of the wreckage and the location of the survivor. Spotting the survivor amid the rough seas, Edwards tossed out three paint markers before realizing he was tossing bags of sand. Having spotted the survivor himself, Sharp told Edwards to "prepare to land" as he lowered the nose of Sikorsky for a landing as close as possible to the survivor.[49] In his words, Sharp admitted, "I made a butchered job of the landing," hitting on top of a wave, descending into the trough below the crest of the next wave and bouncing up again and settling down on a third wave. In Edwards's words, "Hugh hit the water, and I went down in the deck, right down on the floor. He gave it the power and went up again, and about the time I got her, boy he drowned me again. So after the third time [he] stuck to the water."[50]

Sharp's hard landing in the eight- to 10-foot seas damaged the left wing's pontoon. Edwards opened the upper hatch of the fuselage and climbed out on top of the cabin and stood up, holding onto the wing to see above the eight to ten-foot waves. Spotting the survivor, Edwards yelled steering instructions to Sharp above the roar of the engine. Reaching back inside the fuselage for the rope, Edwards discovered all he had was about 10 feet of tie-down rope. Recognizing the hopelessness of trying to use the rope, Edwards instead guided Sharp closer to the survivor. Climbing out onto one of the wing pontoons, Edward reached out to take the hand of the man. The survivor, Lieutenant Cross, let out a "god-awful yell" as Edwards grabbed him, since the salt water and gasoline had burned his skin. Pulling Cross over to him, Edwards managed to get the injured man inside the aircraft. Cold, delirious, and suffering from a broken back, Cross was in rough shape. Edwards took his shirt off and wrapped it around Cross to try and warm him on the back seat of the cabin.[51]

With Cross shivering inside the cabin, Sharp called Edwards's attention to the now-flooded left wing pontoon. The added weight dragged the wing into the water, forcing Sharp to steer into the wind to keep the wings above the waves. By 1905, a patrol from the Fourth Task Force at Parksley arrived on the scene, followed five minutes later by another aircraft from Rehoboth. To counterbalance the flooded pontoon, Edwards climbed out to the right pontoon and sat down on one of the support struts. With a degree of stability restored, a solid engine and functional radio, Sharp informed Virdin and the base of his intentions. Unable to take off, he would taxi the 15 miles to shore. With the Coast Guard now informed of the situation, Sharp aimed for Chincoteague Inlet.[52]

With Cross moaning and raving in agony and Edwards half-naked outside sitting on a strut and fighting off the wind and waves, Sharp began the slow journey to shore. Observing Sharp underway and slowly taxiing westward, Virdin turned for base. During the return journey Virdin and Edmondson learned of Cross's condition and that Shelfus went down with the aircraft. Several hours into the Sikorsky's slow journey westward, Coast Guard boats from Assateague and Ocean City, Maryland, appeared on the horizon. Taking the aircraft in tow, the Coast Guard brought the Sikorsky into Chincoteague Inlet and their station, arriving at 0130 on 22 July. Sharp and Edwards carried Cross to a waiting ambulance, which brought him to the Peninsula General Hospital in Parksley, Virginia, for treatment of his broken back, chemical burns, and exposure. The Sikorsky was pulled up on the beach at Chincoteague, the float drained, and Sharp flew the bird back to Rehoboth on the twenty-second.[53]

After a long stay in the hospital, Cross recovered and served as the inaugural operations officer for Coastal Patrol Base No. 14, Panama City, Florida.[54] Despite considerable searching, Shelfus's remains were never recovered. Only 19 years old at the time of his death, he was ambitious aviator known in his hometown for having flown an airplane under one of the large bridges in downtown Columbus, Ohio. He had enlisted in CAP on 13 March and reported for coastal patrol duty on 1 May. He and his wife had been wed for just over one month, having married on 15 June in what was reported by newspapers as the first Civil Air Patrol wedding. First Lieutenant Charles E. Shelfus became the first CAP member killed in World War II, and the first of 26 personnel killed during coastal patrol operations.[55]

In November, Sharp and Edwards received inaugural CAP Distinguished Service Citations for the rescue of Cross.[56] Shortly after the new year, the two men found themselves summoned to the White House. There on 17 February 1943 in the Oval Office, President Roosevelt awarded both the Air Medal "for meritorious achievement while participating in aerial flight."[57] They thus became the first civilians ever awarded the Air Medal.[58] Maj Gen George E. Stratemeyer, chief of the Air Staff, attended the presentation and lauded Sharp and Edwards as exemplars of "the spirit of the Civil Air Patrol, in cooperation with the Army Air Forces, in the flying of many types of special war missions necessitated by our war effort."[59]

Figure 19. On 17 February 1943, Pres. Franklin D. Roosevelt presented the Air Medal to Maj Hugh R. Sharp Jr., CAP, and Ensign Edmond Edwards, USNR, in the Oval Office for their heroic rescue of 1st Lt Henry Cross. James M. Landis, director of the Office of Civilian Defense, stands to the left. (Photograph courtesy of the Morse Center.)

By the time Sharp and Edwards met Roosevelt, CAP coastal patrol aircrews had located or assisted in the rescue of survivors of Army Air Forces and Navy air crashes, as well as lost Coast Guard and Navy vessels.[60] CAP's increased success in locating survivors owed much to an expansion of the coastal patrol operation to remove any gaps in the nation's inshore coastal air coverage. By late September 1942, CAP boasted 21 bases stretching from nation's northeastern border with Canada to its southeastern border with Mexico.[61] The collective blanket of Army, Navy, Coast Guard, and CAP aircraft, as observed in a 1945 Army Air Forces study, "undoubtedly exercised a determining influence in the enemy's strategic withdrawal," although it remained clear the enemy was not defeated but had "merely concentrated his efforts in other areas."[62]

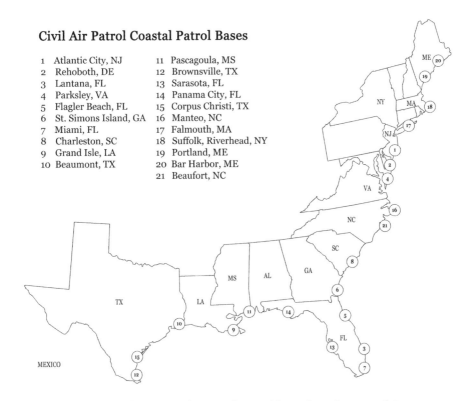

Figure 20. Civil Air Patrol coastal patrol base locations as of September 1942. (Illustration by Maj Erik Koglin, courtesy of the Morse Center.)

For mariners cruising the inshore shipping lanes, added CAP patrols provided lifesaving providence when nature proved the enemy. For the men of Coastal Patrol Base No. 21, Beaufort, North Carolina—the last of the 21 CAP bases—lifesaving patrols began within their first weeks of operation. On 9 November 1942, the 50-year-old, three-mast schooner *Mayfair* encountered serious trouble around noon while sailing 15 miles off Wrightsville Beach. Foundering with seas opening up, the crew of the vessel, Captain Frank C. Sweetman Jr., Elsie V. Sweetman, and Fred S. Sweetman from Brigantine, New Jersey, and Doily T. Willis and Charles W. Willis of Morehead City,

North Carolina, abandoned ship into a small life raft. One crewmember, Y. Z. Newberry Jr. of Morehead City, went down with the ship and drowned. Adrift in cold, rough seas, the survivors huddled aboard their raft for four hours until a CAP dusk patrol happened upon them. Around 1600, 1st Lt John F. Davis of Greensboro and his observer, 2nd Lt Francis W. McComb of Charlotte, spotted the raft and radioed the position to the Coast Guard. A second aircraft arrived, piloted by 1st Lt Alfred C. Kendrick of Gastonia with observer 2nd Lt Herbert O. Crowell of Lenoir, and the two aircraft circled the raft until they were relieved and the survivors were safely aboard a Coast Guard vessel. The Navy did not mention CAP's role in the incident, so the civilian volunteers issued their own statement about the rescue to the press.[63]

Minor publicity problems aside, CAP's coastal patrol growth overwhelmed the supervisory abilities of the I Ground Air Support Command. By 22 June 1942, only 82 of 137 coastal patrol aircraft were equipped with bomb racks as the support command lacked the personnel to handle both Army and CAP needs.[64] To alleviate the situation, on 17 August, I Bomber Command, under First Air Force, took over general supervision, administration, training, and operations of the CAP coastal patrol bases from I Ground Air Support Command. I Bomber Command delegated administrative, training, and operational control for CAP to the I Patrol Force through the 59th and 65th Observation Groups. All CAP coastal patrol operations would be conducted under the supervision of the First Air Force (via I Patrol Force). I Patrol Force would issue instructions defining the areas to be covered, missions performed, and procedures to be followed. These instructions would be followed through by CAP National Headquarters, via Blee. He would also be responsible for the execution of the I Patrol Force instructions governing the organization and administration of the coastal patrols. Establishment, maintenance, and operation of all radio communications between aircraft and the bases remained a CAP matter, but First Air Force would provide communication connections from the coastal patrol bases to the nearest I Bomber Command unit. I Patrol Force would provide liaison officers in an advisory capacity only for periods not to exceed 60 days.[65]

At the end of August, CAP National Headquarters also issued a revised operations directive for all coastal patrol operations. In addition to clarifying the new policy governing the operation of the coastal patrol, all patrol bases were now organized for a maximum authorized strength of 15 pilots and observers, rather than 15 aircraft. Bases could now carry more than 15 aircraft, "determined by operating requirements," although only 15 aircraft would fly on patrol in a given day except upon written authorization from CAP National Headquarters. Bases also received authorization to have one aircraft in operation for "auxiliary service flights" transporting supplies, equipment, or personnel. These aircraft had to be less than 90 horsepower and were forbidden for coastal patrol duty.[66] This shift enabled bases to theoretically always have sufficient aircraft available for all daily patrols, allowing other aircraft to undergo required maintenance or 100-hour inspections.

Concurrent with the shift in organizational control, the coastal patrol's geographic range expanded. On 4 September, CAP National Headquarters requested the removal of the 15-mile patrol limitation for the coastal patrols and a clause substituting in instructions for future operations to "extend such distance offshore as the capabilities of personnel and equipment will permit."[67] I Bomber Command approved the request within a week.[68] CAP aircraft thereafter began to venture upwards of 60 to 100 miles offshore for antisubmarine patrol, convoy escort duty in the shipping lanes, and special missions.[69] As with the arming of the aircraft, the removal of patrol restrictions represented a growing confidence in the proficiency of CAP personnel. Three days later on 7 September, I Patrol Force issued a new mission statement for all CAP coastal patrol units: "To conduct a continuous patrol over coastal shipping lanes during all daylight hours for the purpose of protecting friendly shipping and or locating and reporting enemy submarines, enemy warships, or suspicious surface craft and to take such action as equipment permits in destruction of enemy submarines; to conduct such special missions as are directed by this headquarters."[70]

To further assist the bases in both safety and communication, I Bomber Command began providing teletypewriter exchange services (TWX) and weather service information beginning in September through October. Under the arrangement, the US Weather Bureau would provide the coastal patrol bases with meteorological instrument installations at all bases and weather forecasts for flight operations. First Air Force in turn could provide TWX service for each coastal patrol base for the reception of confidential coastal patrol operational communications and the transmission and reception of weather data. Once complete, bases could receive forecasts and specialized weather advice from the geographically located bureau offices, and the aircraft could receive ciphered weather data in flight. The bases would provide not only ground reports but also reports from aircrew of any significant or unusual weather conditions observed offshore. The latter information aided the bureau's meteorologists, having lost their previous data source of ships at sea owing to war-imposed radio silence.[71]

The reorganization of CAP's coastal patrol operations preceded a much greater effort by the Army to take the strategic offensive against the U-boat. On 15 October 1942, the War Department activated the Army Air Forces Antisubmarine Command (AAFAC).[72] Marshall had informed King about the new command in mid-September, specifying that the operational control of Army aircraft would remain assigned to the Navy, a plan that King openly supported.[73] Arnold received orders on 22 September to organize the command with the resources of I Bomber Command, which had been slated for deactivation.[74] CAP's costal patrol personnel continued to operate under previously issued policies and directives as they moved to the operational control of AAFAC.[75] A culmination of the lessons learned since December 1941, an official Army Air Forces history notes that AAFAC "was able to attack its problems with undivided energy, free from any immediate uncertainty" about its primary mission: to attack hostile submarines anywhere they operated.[76]

Before deactivation, officers from I Bomber Command compiled a report based on inspections of 20 of CAP's 21 coastal patrol bases. As of 3 October, the command reported 1,663 CAP personnel and 365 aircraft, 205 in commission and 172 equipped to carry ordnance. Inspectors typically found high morale and discipline among the base personnel. At Parksley, the inspecting officer reported "excellent housing for headquarters and operations and excellent hangar and repair shop facilities." Atlantic City, New Jersey, and Lantana, Florida, ranked as the best bases of the coastal patrol featuring the most well-trained personnel and excellent facilities. Regarding aircraft armament, the offensive teeth of CAP's effort, inspectors noted a potential for the arming of 85 aircraft to carry 250-pound bombs, and the equipping of a further 13 aircraft with 325-pound depth charges.[77]

Figure 21. Plotting board operator Cpl Louise T. Story of Lenoir, North Carolina, works on the teletypewriter at Coastal Patrol Base No. 16, Manteo. (Photograph courtesy of the Morse Center.)

The AAFAC organization and missions for CAP materialized in November. The command consisted of the 25th and 26th Antisubmarine Wings headquartered in New York and Miami, respectively, to coincide with the Eastern and Gulf Sea Frontiers.[78] CAP coastal patrol bases organized within the sea frontiers were assigned to the respective wing, and the commanding officer of the respective wing exercised operational control of the CAP bases.[79] AAFAC tasked CAP coastal patrol units with a new mission:

> To patrol coastal shipping lanes as directed during daylight hours for the purpose of protecting friendly shipping and of locating and reporting enemy submarines, enemy warships or suspicious surface craft and to take such action as equipment permits in destruction of enemy submarines; to conduct such special antisubmarine missions as are directed by Headquarters Army Air Force Antisubmarine Command.[80]

AAFAC provided CAP with specific patrol and operational guidance far exceeding previous instructions. All coastal patrol bases would maintain two aircraft with combat crews on alert during daylight hours on call for on-command missions. Patrols would now be limited to no more than 60 miles offshore. Escort missions received specific maneuver instructions.⁸¹ In May 1943, AAFAC provided CAP with a refined convoy escort plan developed by the Division of War Research of Columbia University. The mathematically developed plan involved the least amount of navigation for the two-man CAP aircrew but provided complete convoy coverage with light aircraft.⁸² In their totality, the AAFAC instructions sought to standardize and professionalize CAP's antisubmarine warfare capabilities on par with Army and Navy procedures, integrating CAP as equitably as possible with uniformed military operations.⁸³

CAP's shift to AAFAC occurred shortly after the organization commenced operations patrolling the US–Mexican border. This mission owed its existence directly to the coastal patrol subexperiment and the Army recognition and acceptance of CAP's semi-military development. The origins of the southern border effort dated to at least early March, with added engagement in April at the suggestion of Major General Ulio.⁸⁴ In May, Arnold inquired with the commanding generals of the First through Fourth Air Force and the other Army Air Forces commanders about who was already using or could use CAP resources. He noted CAP under favorable conditions could perform a variety of missions, including border observation and patrol.⁸⁵ Having lost the 120th Observation Squadron under his jurisdiction, Gen Walter Krueger, commanding the Southern Defense Command, requested CAP aircraft to patrol the Mexican border on 24 July. From August through September, Maj Harry K. Coffey of Portland, Oregon, worked with Army ground officers to prepare a plan of operations.⁸⁶

With personnel and aircraft from the California Wing and the Gulf Coast coastal patrol bases, the Southern Liaison Patrol commenced operations on 3 October. As later articulated by CAP National Headquarters in December 1942, the mission of the Southern Liaison Patrol was

> to patrol the boundary [border] to prevent any surprise attack on isolated sections of the country by enemy forces or raiding parties; to detect any suspicious activities or the presence of suspicious persons, conveyances, constructions, installations, or directional markings; to prevent the observation by enemy elements of our operating units; to report suspicious aircraft, activity or flights, whether by American aircraft or enemy aircraft; to report possible acts of sabotage visible from the air; to detect any signals which might be directed to enemy units on either side of the border; to cooperate to the utmost with our ground troops on any specified mission or occasion that might arise; and to transport personnel or correspondence between units of the Southern Land Frontier.⁸⁷

The effort operated from two main bases, Liaison Patrol Base No. 1 at Laredo, Texas, and Liaison Patrol Base No. 2 at El Paso. Two subbases at Del Rio and Marfa, Texas, completed the endeavor. Less than a month passed between Coffey's final report to launch the effort and the first missions beginning at Base No. 1. From dawn to dusk, patrols from Brownsville, Texas, to Douglas, Arizona, covered over 1,000 miles of the border flying from 50 to 450 feet above the terrain. CAP National Headquarters Operation Directive no. 32 governing the liaison patrol drew extensively from all the operational lessons learned from the coastal patrol. The table of organization allocated a total of 13 aircraft and 61 personnel for Laredo and 102 personnel and 24 aircraft for El Paso, comprising the Southern Liaison Patrol.[88]

As with the coastal patrol bases, the Southern Liaison Patrol provided the Army with a viable patrol capability when other options were unavailable. If not for CAP, the Army's Southern Land Frontier may have been forced to assign thousands of men to patrol the border on foot or in jeeps, personnel needed elsewhere in the war effort. CAP aerial observer reports soon proved more accurate than those of ground observers, and the Southern Land Frontier found CAP's work essential to border operations. The liaison patrol personnel open carried firearms rather than bombs. Like the coastal patrol, the liaison patrol similarly employed men unable to serve in the military, notably 1st Lt George W. Copping of Van Nuys, California, who flew over 1,000 hours of patrol with two artificial legs. The financial, resource, and equipment challenges the CAP personnel of the liaison patrol encountered mirrored many of those of the coastal patrol bases—not to mention the hazardous flying conditions. But unlike the latter, the former began its work without the skepticism of Army leadership. The civilian coastal patrol effort had laid this foundation for all of CAP.[89]

Notes

Epigraph. Charles R. Berg to Earle L. Johnson, 13 November 1942, BLS, CAP-NAHC.

---

1. Doenitz, *Memoirs*, 221.

2. Wiggins, *Torpedoes in the Gulf*, 22–48, 85–87; Morison, *Battle of the Atlantic*, 137; Doenitz, *Memoirs*, 221; and Blair, *U-Boat War*, 576–83.

3. Morison, *Battle of the Atlantic*, 135–38, 142; Blair, *U-Boat War*, 578; and Craven and Cate, *Plans and Early Operations*, 530.

4. Morison, *Battle of the Atlantic*, 142; "Sub Fight Leadership Taken by Adm. Kauffman," *Miami News*, 4 June 1944, 4; and Blair, *U-Boat War*, 582–83.

5. Craven and Cate, *Plans and Early Operations*, 530; and US Air Force, *Antisubmarine Command*, 18.

6. E. H. Eddy, Headquarters I Ground Air Support Command, to Commanding General, Air Forces, Eastern Defense Command and First Air Force, memorandum, subject: Requirements for Civil Air Patrol Services between 18 June 1942, and 31 December 1942, 22 June 1942, Reel A4064, AFHRA.

7. OCD, CAPNHQ, Operations Orders no. 1, Activation of CAP Coastal Patrols, 30 November 1942, Folder 2, Box 6, ELJ, WRHS; Headquarters, I Ground Air Support Command to Civil Air Patrol Authorities, 9th Task Force, New Orleans, La., Letter of Instructions Number 8, 29 June 1942; Headquarters, I Ground Air Support Command to Civil Air Patrol Authorities, 10th Task Force, Beaumont, Texas, Letter of Instructions Number 9, 29 June 1942; Headquarters, I Ground Air Support Command to Civil Air Patrol Authorities, 11th Task Force, Pascagoula, Miss., Letter of Instructions Number 10, 1 July 1942, Reel 38920, AFHRA; and Craven and Cate, *Plans and Early Operations*, 531.

8. Headquarters I Ground Air Support Command to Civil Air Patrols Authorities, 9th Task Force, New Orleans, La., Letter of Instructions Number 8, 29 June 1942; and Headquarters I Ground Air Support Command to Civil Air Patrol Authorities, Civil Coastal Patrol No. 9, Grand Isle, La., Changes Number 1 to Letter of Instructions Number 8, 11 August 1942, Reel 38920, AFHRA.

9. Speiser, "'Joe'–Submarine Hunter," 3–10, CAP-NAHC; Fandison, interview; and CAPNHQ, untitled document marked "Confidential" listing movement of Coastal Patrol Bases, 15 February 1943, Reel 38920, AFHRA.

10. OCD, *CAP Bulletin* 1, no. 15, 8 May 1942, Folder 6, Box 6, ELJ, WRHS.

11. OCD, *CAP Bulletin* 1, no. 19, 5 June 1942, Folder 6, Box 6, ELJ, WRHS.

12. OCD, *CAP Bulletin* 1, no. 25, 17 July 1942, Folder 6, Box 6, ELJ, WRHS.

13. Earle L. Johnson to Walter Earl Brown, 23 May 1942, BLS, CAP-NAHC. This is one of several examples of this letter in the collection. Johnson was promoted to major on 1 May 1942. Blazich, "Earle L. Johnson," 13.

14. OCD, CAPNHQ, James F. McBroom to All CAP Coastal Patrol Commanders, memorandum, subject: OCD Form no. 630, Application for Assignment of Aircraft to Active Duty, 27 August 1942; and OCD, CAPNHQ, Earle L. Johnson to CAP Members Owning Registered Aircraft, memorandum, subject: Availability of Aircraft for Active Duty, 28 August 1942, BLS, CAP-NAHC.

15. Blair, *U-Boat War*, 588–90; and US Air Force, *Antisubmarine Command*, 10–20.

16. George C. Marshall to Ernest J. King, memorandum, subject: Submarine Sinkings, 19 June 1942, file "A16-3(4)–Overflow–File #2, Warfare Operations–General," Box 260, Headquarters COMINCH, 1942–Secret, A163(3) to A16-3(5), RG38, NARA.

17. Ernest J. King to George C. Marshall, memorandum, subject: Antisubmarine Operations–Protection of Atlantic Shipping, 21 June 1942, file "A16-3(4)–Overflow–File #2, Warfare Operations–General," Box 260, Headquarters COMINCH, 1942–Secret, A163(3) to A16-3(5), RG38, NARA.

18. Blair, *U-Boat War*, 590–95, 691–700; Craven and Cate, *Plans and Early Operations*, 533–35; Roskill, *The Period of Balance*, vol. 2, *The War at Sea*, 102–7; and Doenitz, *Memoirs*, 250–52.

19. Adolphus Andrews to Ernest J. King, memorandum, subject: Submarine Situation during August 1942 and Brief Estimate of Situation, 10 September 1942,

Folder "23. Miscellaneous Analyses by Others," Box 47, Antisubmarine Warfare Analysis and Statistical Section, Series II, Losses in Convoys, RG38, NARA.

20. Ivan Hinderaker to Bernard L. Gladieux, memorandum, subject: Conference with Earle Johnson, National Commander, Civil Air Patrol, 3 August 1942, Folder "OCD Civil Air Patrol," Box 10, Entry 107A, RG51, NARA.

21. Gill Robb Wilson to Reed Landis, memorandum, subject: Uniforms, 31 October 1941, BLS, CAP-NAHC.

22. Helen Rough and Cecile Hamilton to Gill Robb Wilson, memorandum, subject: Uniforms for Civil Air Patrol, 12 November 1941, BLS, CAP-NAHC.

23. Gill Robb Wilson to Reed G. Landis, memorandum, subject: Uniforms, 21 November 1941, Reel 44552, AFHRA.

24. Reed G. Landis to Gill Robb Wilson, memorandum, 21 November 1941, BLS, CAP-NAHC; and Cecile Hamilton to Harry H. Blee, memorandum, subject: Uniforms for Civil Air Patrol, 2 December 1941, Reel 44552, AFHRA.

25. Cecile Hamilton to Gill Robb Wilson, memorandum, subject: Activities on Tuesday, 2 December 1941, Reel 44552, AFHRA; and Walter P. Burn and Helen Jurkops, 1942, Design for an Applique Emblem or Similar Article, US Patent 132,353, filed 11 February 1942, and issued 12 May 1942.

26. Lt Hobson to Maj Clough, memorandum, subject: Meeting, 3 December 1941, Civil Air Patrol Uniforms, 8 December 1941, The Institute of Heraldry, Fort Belvoir, VA (TIOH); and Cecile Hamilton to Harry H. Blee, memorandum, subject: Uniform conference with Quartermaster Corps, Dec. 3, 1941, 3 December 1941, Reel 44552, AFHRA.

27. John F. Curry to Floyd O. Johnson, 19 January 1942, BLS, CAP-NAHC.

28. Emory S. Adams to Fiorello LaGuardia, 2 February 1942, BLS, CAP-NAHC; Clifford L. Corbin to Emory S. Adams, memorandum, subject: Uniform for Civil Air Patrol (Male and Female)–Civilian Defense, 10 January 1942; J. W. Boyer to Quartermaster General, memorandum, subject: Uniforms for Civil Air Patrol, 1st Ind., 14 January 1942; and Emory S. Adams to James M. Landis, 2 February 1942, TIOH.

29. OCD, CAPNHQ, John F. Curry to All Wing Commanders, memorandum, subject: Uniforms and Insignia for Staff Personnel (GM-9), 13 February 1942; OCD, CAPNHQ, Earle L. Johnson to All Wing Commanders, memorandum, subject: Official Officer's Uniform–(Summer) (GM-18), 3 April 1942; and OCD, CAPNHQ, Jack Vilas to All Unit Commanders, memorandum, subject: Sleeve Rank Insignia for CAP Officers (GM-34), 13 June 1942, Binder "Civil Air Patrol–Establishment of Charts, Staff, General Memoranda, Training Memoranda," Box 2, Entry 205, Processed Documents Issued by the OCD–CAP, RG171, NARA.

30. Schlegel, *Virginia on Guard*, 47–48, 201; Harry H. Blee to John F. Curry, 15 January 1942, Folder "507 Civil Air Patrol," Box 111, Entry 10, National Headquarters, General Correspondence, 1940–1942, 502 to 511, RG171, NARA; and L. O. G. to A. E. DuBois, memorandum, 24 February 1942, TIOH. The memorandum only uses initials for the sender.

31. Harry H. Blee to Chief of Staff [George C. Marshall], 9 June 1942, Folder "Civil Air Patrol Historical Documents 1942, Mar–Dec," CAP-NAHC. The exact shade of red CAP selected is Textile Color Card Association Cable No. 14906, known as "Canada Red." Margaret Hayden Rorke to Oscar Smith, 17 August 1942, BLS, CAP-NAHC; and OCD, CAPNHQ, Jack Vilas to All Unit Commanders, memorandum, subject: Replacement of Uniform Shoulder Straps (All Personnel) (GM-49), 27 July 1942, Binder "Civil Air Patrol–Establishment of Charts, Staff, General Memoranda, Training Memoranda," Box 2, Entry 205, Processed Documents Issued by the OCD–CAP, RG171, NARA.

32. James A. Ulio to James M. Landis, 30 June 1942, Folder "Civil Air Patrol Historical Documents 1942, Mar–Dec," CAP-NAHC.

33. War Department, Office of the Quartermaster General, Deputy Director of Production to Resources Division, memorandum, subject: Rank and Grade Insignia for CAP, OCD, with attachments, 20 June 1942, TIOH.

34. *National Defense Act of 1916*, Public Law 64-85, *U.S. Statutes at Large* 39 (1915-1917): 216-17; and Edmund B. Gregory to Civil Air Patrol, Office of Civilian Defense, memorandum, subject: Uniforms, Civil Air Patrol, 1st Indorsement, 16 May 1942, TIOH.

35. Harry H. Blee to Earle L. Johnson, memorandum, subject: Uniforms for Civil Air Patrol, 1 July 1942, BLS, CAP-NAHC; OCD, CAPNHQ, Earle L. Johnson to All Unit Commanders, memorandum, subject: Uniform, Insignia, and Rank (GM-45), 17 July 1942; OCD, CAPNHQ, Earle L. Johnson to All Unit Commanders, memorandum, subject: Qualifications for Appointment with Rank, Commissioned and Non-commissioned Officers (Addendum to GM-45), 17 July 1942, Binder "Civil Air Patrol–Establishment of Charts, Staff, General Memoranda, Training Memoranda," Box 2, Entry 205, Processed Documents Issued by the OCD–CAP, RG171, NARA; and OCD, CAPNHQ, Jack Vilas to All Regional, Wing, and Coastal Patrol Commanders, memorandum, subject: Rank and Grade Designated for Civil Air Patrol Coastal Patrols (GM-50), 29 July 1942, Folder "Civil Air Patrol 15 July 1942," Box 21, Entry 16A, General Correspondence, 1941–May 1945, Civil Air Patrol, RG171, NARA.

36. OCD, *CAP Bulletin* 1, no. 30, 21 August 1942, Folder 6, Box 6, ELJ, WRHS.

37. OCD, *CAP Bulletin* 2, no. 5, 29 January 1943, Folder 6, Box 6, ELJ, WRHS; OCD, CAPNHQ, Jack Vilas to All Unit Commanders, memorandum, subject: Active Duty Emblems (GM-68), 29 December 1942; and OCD, CAPNHQ, Jack Vilas to All Unit Commanders, memorandum, subject: Insignia for Active Duty Volunteers, Merit Awards, and Specialists (GM-39), 2 July 1942, Binder "Civil Air Patrol–Establishment of Charts, Staff, General Memoranda, Training Memoranda," Box 2, Entry 205, Processed Documents Issued by the OCD–CAP, RG171, NARA.

38. OCD, CAP, Operations Orders no. 1, Activation of CAP Coastal Patrols, 30 November 1942, Folder 2, Box 6, ELJ, WRHS. The bases were Brownsville, TX (12th), Tampa, FL (13th), Panama City, FL (14th), Corpus Christi, TX (15th), and Manteo, NC (16th).

39. Morison, *Battle of the Atlantic*, 254–55; Gannon, *Operation Drumbeat*, 387–88; Offley, *Burning Shore*, 240–41; and Murray and Millett, *A War to Be Won*, 251–52.

40. Doenitz, *Memoirs*, 237.

41. Mosley, *Brave Coward Zack*, 46-51; Zack Mosley to John F. Curry, 7 January 1942; Zack Mosley to Earle L. Johnson, 14 May 1942; Earle L. Johnson to Zack Mosley, 25 May 1942; Zack Mosley to Earle L. Johnson, 20 July 1942; Earle L. Johnson to Zack Mosley, 29 July 1942; Zack Mosley to Kendall K. Hoyt, 20 July 1942; Kendall K. Hoyt to Zack Mosley, 31 July 1942; Zack Mosley to Kendall K. Hoyt, 5 August 1942; and Kendall K. Hoyt to Zack Mosley, 9 August 1942, BLS, CAP-NAHC; and Ed Stone, "Zack Mosley, Smilin' Jack's Creator, Makes Home With His Family in Florida," *Miami Herald*, 26 July 1942, 21.

42. Mosley, *Brave Coward Zack*, 52; Zack Mosley to Kendall K. Hoyt, 20 July 1942; Kendall K. Hoyt to Zack Mosley, 31 July 1942; Zack Mosley to Kendall K. Hoyt, 5 August 1942; and Kendall K. Hoyt to Zack Mosley, 9 August 1942, BLS, CAP-NAHC; Elizabeth W. King, "Heroes of Wartime Science and Mercy," *National Geographic* 84, no. 6 (December 1943): 739–40, plate VIII; Sharp, interview; and Boudreau, *CAPCP Base-17*, 1.

43. The insignia for Bases 16 and 18 were featured in the "War Insignia" installments of *Walt Disney's Comics and Stories*. Base 16 opted not to use the Disney design and instead operated without a distinctive insignia. Base 14's insignia was designed and drawn by Walt Disney Productions and used by base personnel but never publicized by Disney. "War Insignia," *Walt Disney's Comics and Stories* 3, no. 7, April 1943; "War Insignia," *Walt Disney's Comics and Stories* 3, no. 9, June 1943; Coastal Patrol Base No. 14, *Buckeye Beacon* 1, no. 4 (18 January 1943): 2; and Vernon Caldwell to Robert E. Church, 10 February 1943, CP16, CAP-NAHC.

44. Keil, "Civil Air Patrol Coastal Patrol No. Three"; and Mosley, *Brave Coward Zack*, 51–54.

45. The aircraft entered a high-speed stall. In the words of Edmond I. Edwards, Cross "pure and simple, he didn't know how to fly, and there was a lot of us down here [who] didn't know how to fly either. He wasn't the only one." Edwards, interview.

46. Carl L. Virdin, memorandum, subject: Crash of Warner Fairchild NC-19144, undated, BLS, CAP-NAHC.

47. Sharp, interview.

48. Frebert, *Delaware Aviation History*, 75, 78. Sikorsky S-39B, NC803W underwent a 14,000-hour restoration and is on display in the New England Air Museum. "Sikorsky S-39B 'Jungle Gym,'" *New England Air Museum*, https://www.neam.org/ac-sikorsky-s39b.php; and Eric Ferreri, "Volunteers Took Battered Plane Under Their Wing," *Hartford Courant*, 1 November 1996, https://www.courant.com/news/connecticut/hc-xpm-1996-11-01-9611010382-story.html.

49. Carl L. Virdin, memorandum, subject: Crash of Warner Fairchild NC-19144, undated, BLS, CAP-NAHC; and Edwards, interview.

50. Sharp, interview; and Edwards, interview.

51. Sharp, interview; and Edwards, interview.

52. Sharp, interview; Edwards, interview; and Carl L. Virdin, memorandum, subject: Crash of Warner Fairchild NC-19144, undated, BLS, CAP-NAHC.

53. Sharp, interview; Edwards, interview; Carl L. Virdin, memorandum, subject: Crash of Warner Fairchild NC-19144, undated, BLS, CAP-NAHC; "CAP Observer Drowns When Plane Crashes," *News Journal* (Wilmington, DE), 22 July 1942, 1; "CAP Observer Drowns, Pilot Rescued at Sea," *Daily Press* (Newport News, VA), 23 July 1942, 5; and "One Drowns, Second Hurt in Shore Plane Crash," *Daily Times* (Salisbury, MD), 23 July 1942, 1.

54. Keefer, *From Maine to Mexico*, 317; and Green, "History of Coastal Patrol Base #14," 11. CAP National Headquarters issued orders for both Cross and Shelfus to transfer to the Panama City base on 20 July 1942. OCD, CAPNHQ, Special Orders no. 84, 20 July 1942, Reel 38913, AFHRA.

55. Documents from Mark Frye shared with author by Seth Hudson via email, 10 August 2016; and assorted news clippings from Ohio Wing scrapbook, 1942–1946, CAP-NAHC.

56. OCD, CAPNHQ, Jack Vilas to All Unit Commanders, memorandum, subject: Citation Order no. 1, 27 November 1942, BLC, CAP-NAHC. Edwards is not listed on the document, but he is wearing the award insignia in the photograph of him receiving the Air Medal from Roosevelt. The omission is presumably because Edwards left CAP and Rehoboth Beach in December 1942 to take a commission with the Naval Reserve as a flight instructor. Edwards, interview.

57. War Department, General Orders no. 1, 4 January 1943, 2; "F.D.R. Decorates 2 CAP Fliers," *Sentinel* (Carlisle, PA), 18 February 1943, 2; and "State CAP Men Given Medals," *News Journal* (Wilmington, DE), 17 February 1943, 1.

58. "Army Sets Precedent with Awards for CAP," *Detroit Free Press*, 28 February 1943, 26; and Spink, "Distinguished Flying Cross," 13.
59. OCD, *CAP Bulletin* 2, no. 8, 19 February 1942, Folder 6, Box 6, ELJ, WRHS.
60. Harry H. Blee to Earle L. Johnson, memorandum, subject: Coastal Patrol Operations, July 23 to 29, 1942, inclusive, 31 July 1942; Harry H. Blee to Earle L. Johnson, memorandum, subject: Coastal Patrol Operations of Civil Air Patrol, September 10 to 16, 1942, inclusive, 17 September 1942; Harry H. Blee to Earle L. Johnson, memorandum, subject: Coastal Patrol Operations of Civil Air Patrol, September 24th to 30th, 1942, inclusive, 2 October 1942; Harry H. Blee to Earle L. Johnson, memorandum, subject: Coastal Patrol Operations of Civil Air Patrol, October 1, to October 7, 1942, inclusive, 8 October 1942; Harry H. Blee to Earle L. Johnson, memorandum, subject: Coastal Patrol Operations of the Civil Air Patrol, December 10 to December 16, 1942, inclusive, 17 December 1942; Harry H. Blee to Earle L. Johnson, memorandum, subject: Coastal Patrol Operations of Civil Air Patrol, January 21 to 27, 1943, inclusive, 29 January 1943; and Harry H. Blee to Earle L. Johnson, memorandum, subject: Coastal Patrol Operations of Civil Air Patrol, February 4 to 10, 1943, inclusive, 11 February 1943, Folder "Civil Air Patrol," Box 7, Entry 233, General Correspondence, Director's Office, Feb. 1942–June 1944, CAP–Budget Estimates, RG171, NARA; and Warner and Grove, *Base Twenty-one*, 68, 74–75.
61. Earle L. Johnson to Henry L. Stimson, 28 July 1942, BLS, CAP-NAHC; OCD, CAPNHQ, Operations Orders no. 1, Activation of CAP Coastal Patrols, 30 November 1942, Folder 2, Box 6, ELJ, WRHS. The final bases were located at Riverhead, NY (17th); Falmouth, MA (18th); Portland (19th) and Bar Harbor (20th), ME; and Beaufort, NC (21st). CAP task forces were redesignated as "Civil Coastal Patrol #" on 26 July 1942 and then redesignated again as "Civil Air Patrol Coastal Patrols" on 20 August 1942. Edward J. Culleton to all concerned, memorandum, subject: Redesignation of Civil Air Patrol Units on Coastal Patrol (General Memorandum Number 1), 26 July 1942; and OCD, CAPNHQ, Harry H. Blee Station List no. 1–CAP Coastal Patrols, 20 August 1942, Reel A4064, AFHRA.
62. US Air Force, *Antisubmarine Command*, 36.
63. "Tar Heel CAP Fliers Aid in Ocean Rescue," *News and Observer* (Raleigh, NC), 16 November 1942, 8; "Schooner Sinks Just Off Carolina," *News Journal* (Wilmington, DE), 11 November 1942, 11; "Jerseyans Saved as Schooner Sinks," *Courier-News* (Bridgewater, NJ), 11 November 1942, 17; "CAP Sights Raft and Saves Five," *Rocky Mount* (NC) *Telegram*, 13 November 1942, 9; Warner and Grove, *Base Twenty-one*, 61–62; and War Diary, Eastern Sea Frontier, July 1943, chap. 4, 16–18, and attachments, NARA (via Fold3).
64. E. H. Eddy, Headquarters I Ground Air Support Command, to Commanding General, Air Forces, Eastern Defense Command and First Air Force, memorandum, subject: Requirements for Civil Air Patrol Services between 18 June 1942, and 31 December 1942, 22 June 1942, Reel A4064, AFHRA.
65. Charles L. Carlson to Henry H. Arnold, memorandum, subject: Civil Air Patrol, 20 July 1942, Reel A4064; Louis E. Boutwell to Chief of Staff, Headquarters, I Bomber Command, memorandum, subject: Plan of Organization and Operation, I Patrol Task Force, 15 August 1942; Headquarters, Air Forces, Eastern Defense Command and First Air Force, E. E. Glenn, General Order Number 59, 17 August 1942, Reel A4050; Robert W. Harper to Commanding General, Headquarters Air Forces, Eastern Defense Command and First Air Force, memorandum, subject: Amendment of Policy on Operational Control of the Civil Air Patrol, 25 August 1942; Charles L. Carlson to Westside T. Larson, memorandum, subject: Plans for Organization and Training, and Enclosure no. 1, Plan for General Supervision of C.C.P.

Units and for Control, Organization and Training of Observation Units, 18 August 1942; Louis E. Boutwell to all C.C.P. Units, memorandum, subject: Change of Command, 22 August 1942; Earl J. Nesbitt to Commanding General, Eastern Defense Command and First Air Force, Mitchel Field, New York, memorandum, subject: Letter of Transmittal, 17 September 1942 with attachment, Plan for General Supervision of C.A.P. (Coastal Patrol) Units Administration, Operations and Training of the 59th and 65th Observation Groups, Reel A4064, AFHRA; and telegram from 65th Observation Group, Langley Field, Virginia to CAP Coastal Patrol Base No. 16, Manteo, North Carolina, 15 August 1942, CP16, CAP-NAHC.

66. OCD, CAPNHQ, Operations Directive no. 23A, CAP Coastal Patrols, 26 August 1942, AWS, CAP-NAHC.

67. Earle L. Johnson to Louis E. Boutwell, memorandum, subject: Removal of 15-Mile Limitation–CAP Coastal Patrols, 4 September 1942, Reel A4064, AFHRA.

68. Antisubmarine Command Historical Section, "CAP History of Operations (First Narrative)," 12 October 1943, 3, Reel A4057; Louis E. Boutwell to James E. Chaney, memorandum, subject: Removal of 15-Mile Limitation–CAP Coastal Patrols, 1st Indorsement, 5 September 1942; and Charles L. Carlson to Westside T. Larson, attention I Patrol Force, memorandum, subject: Removal of 15-Mile Limitation–CAP Coastal Patrols, 2nd Indorsement, 10 September 1942, Reel A4064, AFHRA.

69. House Committee on Military Affairs, *Civil Air Patrol: Hearings on H.R. 1941 and H.R. 2149*, 79th Cong., 1st sess., 1945, 3, 6, 23.

70. Louis H. Boutwell, Headquarters I Bomber Command, I Patrol Force, to all Civil Air Patrol Coastal Patrol Units (thru CAPCP National Headquarters, Washington, D.C.), Letter of Instructions Number 1, 7 September 1942, Binder "Legal Status, Administrative Concepts, and Relationships of the Civil Air Patrol, 1941 to 1949," CAP-NAHC.

71. Harry H. Blee to Ralph L. Higgs, memorandum, subject: Visit to Coastal Patrol Task Forces, 27 July 1942, Reel 38919; Charles L. Carlson to Westside T. Larson, memorandum, subject: TWX services for Civil Air Patrol Coastal Patrol Units, 5 October 1942; Louis E. Boutwell to Headquarters Air Forces, Eastern Defense Command and First Air Force, Office of the Air Force Commander, memorandum, subject: 1st Ind., 12 October 1942; Harry H. Blee to Louis E. Boutwell, memorandum, subject: TWX Teletype and Weather Service for CAP Coastal Patrol Bases, 23 October 1942; OCD, CAPNHQ, Coastal Patrol Circular no. 8, Weather Service–CAP Coastal Patrols, 1 October 1942; and Francis W. Reichelderfer to Harry H. Blee, 21 September 1942, Reel A4064, AFHRA. By March 1943, seven coastal patrol bases still lacked TWX service (Bases 1, 7, 13, 16, 18, 20, and 21). Harry H. Blee to the New York Telephone Company through Headquarters, Army Air Forces Antisubmarine Command, memorandum, subject: TWX Teletype Installations–CAP Coastal Patrol Bases, 6 March 1943, BLS, CAP-NAHC.

72. D. T. Sapp, War Department, Adjutant General's Office, to Commanding General, First Air Force, memorandum, subject: Constitution, Activation, Inactivation and Reassignment of Certain Army Air Force Units, 13 October 1942, Box 3, Entry 117, Office, Special Consultant to the Secretary of War, RG107, NARA; Craven and Cate, *Plans and Early Operations*, 552–53; Headquarters, Air Forces, Eastern Defense Command and First Air Force, General Order no. 84, 15 October 1942, Reel A4063, AFHRA; and Schoenfeld, *Stalking the U-Boat*, 2.

73. George C. Marshall to Ernest J. King, memorandum, subject: Formation of First Anti-Submarine Army Air Command, 14 September 1942; Ernest J. King to George C. Marshall, memorandum, subject: Formation of First Anti-Submarine Army Air Command, 17 September 1942, Box 3, Entry 117, Office, Special Consul-

tant to the Secretary of War, RG107; and Fred C. Milner, Headquarters, Army Air Force Antisubmarine Command, to Commanding General, Army Air Forces Antisubmarine Command, memorandum, subject: Army Air Forces Antisubmarine Command, 28 December 1942, file "A16-3(9)–Submarine & Anti-Submarine Warfare, 1943. File #1," Box 672, Headquarters COMINCH, 1943–Secret, A16-3(9), RG38, NARA.

74. Joseph T. McNarney to Henry H. Arnold, memorandum, subject: Formation of First Anti-Submarine Army Air Command, 22 September 1942, Box 3, Entry 117, Office, Special Consultant to the Secretary of War, RG107, NARA.

75. Louis E. Boutwell, Headquarters A.A.F. Anti-Submarine Command, I Patrol Force, to Earle L. Johnson, memorandum, subject: Change of Command and Command Channels, 19 October 1942, Reel A4064, AFHRA.

76. Craven and Cate, *Europe: Torch to Pointblank*, 378.

77. Louis E. Boutwell to Westside T. Larson, memorandum, subject: Report of Inspection: CAP Units, 3 October 1942, Reel A4064, AFHRA.

78. Craven and Cate, *Europe: Torch to Pointblank*, 378; Maurer, *Combat Units*, 388–89; and Louis E. Boutwell to Earle L. Johnson, memorandum, subject: Change of Command and Command Channels, 19 October 1942, Reel A4064, AFHRA.

79. Headquarters, Army Air Forces Antisubmarine Command, Operational Instructions, Annex no. 1, Letter of Instructions Number 1, 27 November 1942, Reel A4063, AFHRA.

80. Headquarters, Army Air Force Antisubmarine Command, G.A. McHenry to Commanding Officer, 25th Wing AAF Antisubmarine Command, Commanding Officer, 26th Wing AAF Antisubmarine Command, and All CAP Coastal Patrol Units (Thru CAP National Headquarters), Letter of Instructions Number 1, 27 November 1942, Reel A4063, AFHRA.

81. Headquarters, Army Air Forces Antisubmarine Command, Operational Instructions, Annex no. 1, Letter of Instructions Number 1, 27 November 1942, Reel A4063, AFHRA.

82. G. A. McHenry to Commanding Officer, 25th Antisubmarine Wing, Commanding Officer, 26th Antisubmarine Wing, All CAPCP Units (Through CAP National Headquarters), Letter of Instructions Number 1E, 12 May 1943; and Division of War Research, Columbia University, Memorandum 29–Convoy Escort Plans for the Civil Air Patrol, 28 April 1943, Reel A4063, AFHRA.

83. Headquarters, Army Air Forces Antisubmarine Command, Operational Instructions, Annex no. 1; Signal, Annex no. 2; Safety Precautions, Annex no. 3; Reports, Annex no. 4, Letter of Instructions Number 1, 27 November 1942, CAP-NAHC.

84. A "Report on Civil Air Patrol Missions" dated 6 March 1942 mentions "Border Patrol on Mexican Border to be started when funds are available." In a confidential letter of 9 April 1942 to Ivan H. Hinderaker at the Bureau of the Budget, Johnson mentions "it is anticipated that we will relieve the Army planes which are now patrolling the Mexican Border with pilots and planes of the Civil Air Patrol." An enclosed outline for ten CAP Mexican Border Patrol bases lists a total estimates monthly cost of $110,000. Document, "Report on Civil Air Patrol Missions," 6 March 1942; and Earle L. Johnson to Ivan H. Hinderaker, 9 April 1942, Folder "OCD Civil Air Patrol," Box 10, Entry 107A, RG51, NARA. See also chapter 4.

85. William W. Dick to Commanding General, First through Fourth Air Force, Air Service Command, Materiel Command, Ferrying Command, Flying Training Command, Transport Command, Air Force Proving Ground Command, memorandum, subject: Civil Air Patrol, 19 May 1942, Folder "SAS 324.3–Civil Air Patrol," Box 93, HHA, LOC. The same memorandum is also found dated 30 May 1942. William

W. Dick to Commanding General, First through Fourth Air Force, Air Service Command, Materiel Command, Ferrying Command, Flying Training Command, Transport Command, Air Force Proving Ground Command, memorandum, subject: Civil Air Patrol, 30 May 1942, Reel A4064, AFHRA.

86. Ragsdale, *Wings Over the Mexican Border*, 215–18; Innis P. Swift to Walter Krueger, memorandum, subject: Border Operations, Civil Air Patrol, 6 July 1942; Wilson Davis to Henry H. Arnold, memorandum, subject: Request for Civil Air Patrol Border Patrol, 24 July 1942; Harry K. Coffey to Bertrand Rhine, memorandum, subject: Personnel and Equipment for the Civil Air Patrol, Southern Frontier, Liaison Patrol, at El Paso, Texas, 23 September, 1942; William C. Crane to Harry K. Coffey, 21 September 1942; and Harry K. Coffey, report, subject: Mexican Border Patrol by Civil Air Patrol, 10 September 1942, Reel 38918, AFHRA.

87. OCD, CAPNHQ, CAP Coastal Patrol and Liaison Patrols, 23 December 1942, BLS, CAP-NAHC.

88. OCD, CAPNHQ, Operations Directive no. 32, CAP Southern Frontier Liaison Patrol, 10 November 1942; OCD, CAPNHQ, Operations Directive no. 32, Change no. 3, CAP Southern Liaison Patrol, 8 December 1942, AWS, CAP-NAHC; and Ragsdale, *Wings Over the Mexican Border*, 220.

89. War Department, Headquarters of the Army Air Forces, Dudley M. Outcalt to Air Inspector, memorandum, subject: Survey of the Civil Air Patrol, 8 March 1944, 37–39, Folder 3, Box 5, ELJ, WRHS; Charles B. Rich, Narrated History of Civil Air Patrol Liaison Patrol Base 2, at El Paso, Texas, Reel 38910, AFHRA; OCD, CAPNHQ, Operations Directive no. 32, CAP Southern Frontier Liaison Patrol, 10 November 1942, AWS, CAP-NAHC; and Ragsdale, *Wings Over the Mexican Border*, 220–21, 236–37.

Chapter 6

Challenges and Transitions

The CAP Coastal Patrol carried into their job the unique fellowship and enthusiasm which characterizes amateur aviation. Truly, they were the modern counterpart of the Revolutionary volunteers who flocked to the defense of their homeland in time of emergency with no thought of pay or draft.

—1st Lt E. H. Johnson, 13 September 1943

The renewed Army Air Forces antisubmarine effort in the fall of 1942 coincided with a harder look at CAP's coastal patrol operations. While establishing 21 coastal patrol bases was an impressive accomplishment for such a young organization, the effort strained CAP's available resources to meet its mission requirements and keep its diverse fleet of aircraft operational.[1] Aircraft maintenance at the earlier bases varied in quality, but units at newer bases faced poor or nonexistent facilities, trouble in acquiring parts, and insufficient equipment and personnel for aircraft repair and upkeep.

Despite the guidance provided by CAP National Headquarters, the facilities at the coastal patrol bases represented a constant challenge. Several bases had to relocate operations due to military or commercial necessity for the available facilities or difficulties in fulfilling assigned patrol areas.[2] While some coastal patrol units commenced operations at paved airfields with at least some buildings (in varied condition) for maintenance and personnel, other units confronted considerable logistical challenges. At Grand Isle, Louisiana, the personnel of Base No. 9 staked down their aircraft a mere 150 feet from the beach because they initially had neither a hangar nor the means to construct one. A herd of approximately 100 head of cattle freely roamed the island, and the base could not obtain barbed wire to fence off runways or parking areas where the cattle would eat the fabric off the aircraft.[3] Base No. 14, Panama City, Florida, solved its housing issues by obtaining permission to take buildings from the former Lynn Haven Civilian Conservation Corps camp and move them to the grass airfield with the help of Army equipment borrowed from Tyndall Field.[4]

The two bases along the North Carolina coast both began operations in what was essentially marshland. Base No. 16 in Manteo initially operated off a grass field "hacked out of a swamp," with mosquitos so thick a CAA inspector said the insects covered the base "in clouds."[5] The unit later received permission to move to Naval Auxiliary Air Station Manteo in late October 1942, where the CAP men borrowed a disused sawmill to make lumber from the trees the Navy felled to clear land for the field. In between flight operations, base personnel used the green timber to construct an operations building and hangar.[6] At Base No. 21 in Beaufort, crews flew off fields carved out from marsh grass. Bowing to the sheer persistence of the base commander, Maj Frank E. Dawson of Charlotte, the state civilian defense agency submitted a War Public Works application to the Federal Works Agency to pave the runways, grade the taxiways, and build proper drainage systems. Beaufort transformed from tidal marsh fields into a professional airport by July 1943, the only CAP coastal patrol base to receive such improvement independent of War Department assistance.[7]

Base No. 20 at Bar Harbor, Maine, faced perhaps the most difficult of situations. In the frigid predawn hours of 10 December 1942, an oil heater exploded in the radio room. The conflagration destroyed the joint operations and administration buildings, and the base lost its radio equipment, office equipment, all survival equipment, records, and possessions of many of the personnel. Down but not defeated, base commander Capt James B. King rallied his people and with the support of the towns of Bar Harbor and Ellsworth the base, phoenix-like, reemerged from the ashes in early 1943.[8]

When the coastal patrol effort began to expand in summer 1942, CAP competed with the growing Army Air Forces for enough CAA—certified Airframe and Engine (A&E) mechanics. Base commanders eventually received guidance on the impossibility of assigning more than one A&E-certified mechanic per task force. With an allotment of five mechanics per task force, this meant the remaining four mechanics to would be uncertified but "qualified to do the work under the direction of a certificated mechanic."[9] Competition for certified A&E mechanics among the Army Air Forces, CAP, and other agencies grew to such an extent that CAP National Headquarters issued orders not to recruit mechanics already in the employ of an agency of the Army Air Forces.[10] On several occasions, newly certified A&E mechanics found themselves assigned to coastal patrol bases and bringing whatever tools and spare parts they owned to outfit the repair shops.[11] The certified mechanics often found themselves working long hours overseeing inexperienced personnel with insufficient resources to effect repairs.[12]

Figure 22. Mechanics work on a Fairchild 24 at Coastal Patrol Base No. 6, St. Simons Island, Georgia. (Photograph courtesy of Brooks W. Lovelace Jr. via the Morse Center.)

Searching for spare parts to service an array of airframes and engines proved almost as difficult as spotting submarines. Bases frequently waited from four to six weeks for reimbursement vouchers to arrive from the Treasury Department. Until the funds arrived, personal bank accounts had to suffice to cover repairs.[13] Creativity and interpersonal connections proved invaluable in keeping planes operational when proper supply channels failed to produce results.[14] At Base No. 7 in Miami, mechanics cut the bottoms off beer cans to make a substitute for a rubber wind cone for radio antennae.[15] On Grand Isle, Louisiana, at Base 9, mechanics made baffles out of an old Coca-Cola sign, exhaust stacks from oil drums, aircraft covering out of bedsheets and pillowcases, and antenna weights from toilet plungers. Aircraft cannibalization at the base provided occasional solutions but posed new problems when an aircraft might be spotted sporting two N-numbers, one on each wing.[16] At Coastal Patrol Base No. 17, Suffolk, Riverhead, New York, the engineering department under Capt

Warren E. Moody of Winston-Salem ingeniously converted a Fairchild 24 into a bombing trainer, rigging up an electrically actuated bomb release that helped pilots improve their accuracy.[17] The most heralded example of CAP creative aircraft maintenance is the work of Capt Everett M. Smith, engineering officer and chief mechanic at Rehoboth. Smith and his mechanics improvised with automobile parts and tools for aircraft repairs and safety modifications, establishing a reputation among the base pilots who claimed to fly "by the grace of God and Smitty."[18] Smith's work eventually eased as the base attempted to standardize to only Fairchild 24 aircraft, enabling Rehoboth to perform complete overhauls.[19]

Figure 23. Remains of a crashed aircraft from the Second Task Force, Rehoboth, Delaware, after recovery from the Atlantic. (Photograph courtesy of Henry E. Phipps via the Morse Center.)

From the early months of the coastal patrol effort, the Civil Aeronautics Board conducted inspections of all the CAP bases. Inspectors reported on the airfield, personnel, aircraft, operations, and maintenance. Base maintenance and supply varied. Inspector George M. Keightley, investigating conditions at Base No. 8 in Charleston, South Carolina, found the unit in possession of a "well-equipped maintenance and overhaul department, which was better supplied with motor parts and supplied than any base investigated."[20] By comparison, investigator Julian R. Wagy found Base No. 21 in Beaufort severely lacking. "Considering the lack of proper equipment and the limited facilities available," commented Wagy, "it is quite remarkable that this base is able to carry on operations at all." He further admitted in his report how "the hazardous nature of the operations carried out by the C.A.P. pilots and observers in a very patriotic endeavor causes the C.A.A. Inspector to overlook violations of certain Civil Air Regulations which would not be tolerated in any other type of civil flying."[21]

Figure 24. Personnel of Coastal Patrol Base No. 10, Beaumont, Texas, replacing the fabric covering on what appears to be a Fairchild Model 24. (Photograph courtesy of the Morse Center.)

As early as August 1942, pilot and airframe fatigue caught CAA notice.[22] One investigator documented pilots at bases in Texas recording 53:55 hours and 41:15 hours of patrol time over 12 and 7 days, respectfully.[23] Without any formal regulations on crew rest, CAP aircrews flew an impressive 82,829 patrol hours by October.[24] Most likely as a result of the CAA findings, CAP National Headquarters encouraged coastal patrol commanding officers to appoint flight surgeons for semimonthly inspection of the health and fitness of the base personnel.[25] Furthermore, CAP National Headquarters established a required one day of rest per week for coastal patrol aircrew members, modifying this requirement in November with an additional two days of rest per month.[26] The headquarters also issued a directive for all bases to develop and install checklists in every cockpit for pilots to consult, from engine start to landing and placing safeties on any installed ordnance.[27]

The hastily designed and installed bomb racks also received increased CAA investigation in August to ascertain the integrity of aircraft fuselages. The month prior, Stinson Aircraft had reached out to CAP National Headquarters upon learning of "heavy equipment attached to the lower right fuselage member" on its Stinson Voyager 10A and stated how the fuselage structure would not meet CAA requirements without reinforcement.[28] For some airframes, the Army's solution consisted of using two clamps to fix a bomb rack to the lower fuselage longerons just aft of the passenger cabin door; for others, additional supports were welded to support the load.[29] The addition of 125 pounds (bomb plus rack) invariably increased stress to the tubular frame member, not to mention the added maneuvering stress from steep banks and turns. CAA investigator W. Edmund Koneczny doubted whether the Army Air Forces would find the arming of the civilian aircraft acceptable without structural reinforcement if the CAA or manufacturers provided stress analyses.[30] It is not clear if the analyses occurred, but pilot checklists included inspection of the bomb rack (if installed) prior to engine startup.[31]

In fall 1942, however, the CAP coastal patrol armament situation proved disappointing. Several years before, the British had recognized that a 100-pound bomb, even with a direct hit—difficult with even the best of bombsights—did not guarantee the sinking of a U-boat unless the pressure hull was breached. As the US Navy noted, safety considerations required CAP to drop 100-pound demolition bombs at appreciable altitudes, "which precludes any consistent accuracy."[32] Furthermore, larger, Torpex-filled aerial depth charges proved the ideal weapon.[33] At the time of AAFAC's establishment, less than half of CAP coastal patrol aircraft had bomb shackles installed capable of carrying a single AN-M30 100-pound general purpose demolition bomb, a pair or group of three AN-M30s, or, in far smaller numbers, a single AN-M57 250-pound demolition bomb or Mk 17 325-pound depth bomb.[34] AAFAC did report that a minimum charge of 30 pounds of TNT was the "smallest that with reasonable assurance will afflict lethal damage in direct contact." For larger ordnance, like the Mk 17, a bomb dropped within a 17- to 25-foot radius of a submarine's pressure hull would be lethal. Ergo, a small bomb's lethal radius equated to a contact hit, whereas a large bomb gave more variability for a kill.³⁵

Bombsights and training for bomb runs in turn would be required to increase the probability of accurate attacks— but less than half the CAP coastal patrol aircraft were equipped with bombsights, even the simple sights created by CAP members or Army personnel.³⁶ Actual dropping of bombs varied from base to base as the bombsights and release mechanisms varied. Such variance necessitated practice and familiarization for pilots and observers with each aircraft's particular bomb release mechanism.³⁷ Nonetheless, the primitive equipment and limited formal training did not deter CAP aircrews from attacking when opportunity allowed, with 70 bombs expended in 51 attacks by 14 October 1942.³⁸

Armed or not, the low, slow, brightly colored CAP aircraft on patrol boosted the spirits of the merchantmen below them. "I would say that the morale of the seamen on board my ship was greatly improved by having airplanes flying over the vessel," wrote Frederick Lyall, master of the tanker MS *Pennsylvania Sun*.³⁹ Jacob Pypelink, master of the tanker MS *Sun*, shared similar sentiments, noting, "Even though they [CAP] are unable to attack a submarine, we felt that if there was one [aircraft] in the vicinity they would spot it, and report it to us and the Navy in time to prevent its attacking us."⁴⁰ Such sentiments did much to boost the spirits of the CAP coastal patrol personnel. However, the pace of operations and material limitations at the bases remained problematic.

Lifesaving equipment remained a serious problem for CAP in the fall of 1942. Men flew with store-bought duck hunting life vests, secondhand merchantman kapok life jackets, or improvised flotation gear made from vehicle inner tubes and sea bags copied from Prohibition-era bootleggers. The latter device, the "barracuda bucket" (or bag), intended to provide a CAP member with protection from sharks and other marine life in lieu of a life raft.⁴¹ The life jackets provided by the Coast Guard for personnel at Base No. 21, Beaufort, North Carolina, were removed from the bodies of dead merchantmen and reeked of fuel oil, making flying in them unpleasant; they also swelled upon contact with water, making extrication from a ditched aircraft difficult at best.

Telegrams from angry CAP base commanders and quiet assistance from sympathetic military aviators resulted in "Mae West" life vests being issued at several bases.⁴² The need for life rafts proved even more critical. Having them from the onset of CAP coastal patrol might have saved the lives of a number of downed aircrews, with at least seven members dying from exposure.[43] A survey of the bases that fall revealed over half of them lacked life rafts.[44] In late November, CAP National Headquarters advised all coastal patrol flight personnel "who are not already proficient in the art of swimming" to receive instruction.[45] Adding to the difficulty was locating the survivors in the water. Without dye markers on life vests, personnel at Base No. 7 at Miami created 15' by 15" floating orange markers attached to the backs of aircrew to use as markers in the event of a water ditching.[46]

By December, lifesaving equipment began to be issued with some regularity. As of March 1943 all 21 bases provided one-man life rafts to equip both pilots and observers[47]; they came too late for, however, 1st Lts Julian L. Cooper of Nashville, North Carolina, and Frank M. Cook of Concord, North Carolina. When their aircraft from Base No. 16 at Manteo went down off Pea Island, North Carolina, in the afternoon of 21 December 1942, both men exited the aircraft and inflated their life vests, but the winter ocean soon brought on hypothermia. The Coast Guard tried to reach the men in surf boats but the rough seas made the task impossible. Alive at dusk, the men were gone at dawn. Cook's remains were recovered in March off Cape Lookout. Cooper was never found.[48]

Life rafts, life vests, and the addition of "zoot suits"—early rubber survival suits—gave the coastal patrol bases a needed morale boost. The zoot suits kept an aircrew member dry and warm in the event of a water landing, but they did not provide comfort for any wearer while sitting for hours in a cramped cockpit. The yearbook for Base No. 19 at Portland, Maine, "affectionately" described the zoot suit: "A more miserable cold weather garment was never invented, but orders are orders and we wore them much to the disgust of all concerned."[49] In April 1943, CAP National Headquarters introduced a new award for those members who survived a water landing. These men, having earned web feet, became members of the "Duck Club," a CAP equivalent to the Army Air Forces' Caterpillar Club for those men who "hit the silk" and parachuted safely to earth. Men wore the insignia, a red duck on a field of blue, on their left breast uniform pocket. At the conclusion of the coastal patrol effort, 112 men joined the Duck Club, with Lake Charles, Louisiana, native 1st Lt Louis J. DiCarlo of Base No. 9 surviving two dunkings.[50]

Figure 25. Aircrew from Coastal Patrol Base No. 17, Suffolk, Riverhead, New York, receive a mission briefing while wearing Army Air Forces–issue Type B-4 pneumatic life vests. (Photograph courtesy of the Morse Center.)

Personnel matters represented another issue affecting CAP coastal patrol bases. From the onset of the war, CAP service did not preclude Selective Service. State and local draft boards soon began inducting CAP coastal patrol's most critical personnel, notably pilots and mechanics. As early as mid-July 1942, Johnson had confidentially advised all task force base commanders to work with local draft boards as necessary to retain pilots and mechanics and, if unsuccessful, to contact CAP National Headquarters for assistance.[51] Later that month, he wrote directly to Secretary Stimson asking for some classification to obtain draft deferments for the coastal patrol personnel.[52] His inquiry found a solution by August with Arnold's authorization for the enlistment of CAP coastal patrol pilots and mechanics into the Air Corps Enlisted Reserve (ACER). Any CAP pilot or mechanic who enlisted in ACER at the grade of private would be placed on inactive status until the CAP coastal patrol effort terminated or the individual was dismissed or resigned from the effort, whereby they would be called to active duty in the Army Air Forces.[53]

Only CAP coastal patrol pilots and mechanics could enlist into ACER. While not a deferment, ACER provided a means to retain critical personnel without violating Selective Service obligations. CAP National Headquarters requested that base commanders not urge personnel to enter ACER or submit names of individuals undesirable for duration assignments. Individual members would retain their choice to either enter ACER or work with their local draft board.[54] After 6 December, the War Department ceased accepting further applications from men between the ages of 18 and 38. Of the 442 pilots and 369 mechanics on coastal or liaison patrol duty, only 174—21 percent—entered ACER. CAP National Headquarters estimated 50 percent of pilots and 10 percent of mechanics were over the age of 38 or had physical disabilities making them ineligible for Selective Service.[55]

The matter of replacement parts and priority ratings for the coastal patrols appears to have intensified as fall approached. The Army had agreed in early July 1942 that the coastal patrol bases would receive priority ratings to enable them to secure necessary supplies and replacement aircraft parts; CAP National Headquarters reported having such satisfactory arrangements in place.[56] At that time, 12 coastal patrol bases were operational, four having only operated for a month or less.[57] By October, however, with 21 bases operational and the Army in final preparations for landings in Axis-controlled North Africa, CAP's increased demands outpaced those of the military. Stinson Aircraft approached National Headquarters again about the need to coordinate parts orders, having seen the coastal patrol base supply officers ordering an array of parts, with and without priorities, and without any coordination. Floyd O. Johnson, Stinson's service manager, thought organizing a CAP parts depot could allay unnecessary confusion and delays while permitting the intelligent manufacture of parts to meet CAP demand. He recommended CAP issue a bulletin regarding priority certificates issued by base commanders explaining how Stinson could not ship parts unless provided with complete A-1-A or better priority certificates.[58] Blee, in reply to Johnson, noted the headquarters being "much interested in the suggestions you have to offer on this important subject" and welcomed a visit to discuss the matter directly in Washington.[59]

It is unclear if Floyd Johnson visited with Blee to discuss the parts situation and priority certification, but CAP may have acted because of his suggestions. Around 21 October, Earle Johnson presented a Certificate of Military Necessity with accompanying paper to Arnold. Johnson's certificate and attachment of concerned aircraft and table of operations (6 March–14 October) was addressed to the War Production Board, Joint Army-Navy Munitions Board, and Commanding General, Air Service Command. The certificate stated a military necessity existed for the procurement of repair and replacement additions, accessories, and instruments, as well as radio-controlling equipment for CAP-operated aircraft and engines conducting official missions for the Army Air Forces. Headquarters, Army Air Forces A-3 (Operations) and A-4 (Logistics) did not concur with the military necessity argument, believing that if Arnold signed the certificate this would "be the opening wedge for later furnishing both airplanes and engines" to CAP.[60]

CAP's certificate package moved within the various offices of Headquarters Army Air Forces for additional input. First on the list was Brig Gen Eugene L. Eubank, Director of Bombardment. His directorate concluded CAP in its present status and aircraft could not fulfill the military requirements for its existence. If CAP were to be retained, it "should be militarized and equipped with satisfactory aircraft" but not furnished with replacement aircraft and spare parts "while under its present status." Eubank forwarded the package to Maj Gen Thomas J. Hanley, assistant chief of the Air Staff for Materiel, who concurred with the recommendations but included a statement that by not signing the Certificate of Military Necessity, CAP would gradually disappear through attrition of its aircraft. Before then, however, AAFAC would be strong enough to accomplish its mission alone. Information in hand, Arnold approved all recommendations. Johnson's effort was dead. With CAP's need for spare parts and maintenance facilities in direct competition with similar needs of combat units, the decision by Headquarters Army Air Forces was direct and objective.[61]

On 7 November, Irving H. Taylor, who back in January had proffered the idea of using CAP for coastal patrol work with Curry, reached out to Earle Johnson about the matter of spare parts. He recommended CAP prepare estimates for repair parts and then file the necessary paperwork with the Office of Civilian Supply so manufacturers could produce and deliver spare parts for CAP by 1 July 1943.

Until then, he seconded Stinson Aircraft's suggestion that CAP obtain at least an A-1-A priority. "Now, even with a carefully arranged, authorized schedule of C.A.P. spares production after July 1, 1943," wrote Taylor, "your people are not going to get spares quickly unless you put the manufacturers in a position to produce economic runs and actually turn out the stuff for the account of the Army, Navy, or C.A.P. and warehouse it at conveniently located points." Spare parts and sub-assemblies for 500 coastal patrol aircraft would cost approximately $300,000.[62]

Having failed to secure the Certificate of Military Necessity and the A-1-A priority from Arnold, CAP National Headquarters found itself in a dire situation. When only a few bases were in operation, parts could be secured from existing inventories or at worse, other aircraft could be located to fly when others remained grounded for want of maintenance. With 21 bases flying almost 500 aircraft of varied engines and manufacturers, the existing informal arrangement simply collapsed under the strain of demand. Sympathetic military personnel assisted individual bases where possible, repairing dangerously worn-out aircraft and in several instances overhauling and repainting worn-out civilian aircraft in Navy three-tone camouflage.[63]

On 10 November, Johnson indicated to Headquarters, Army Air Forces his intention to immediately dissolve CAP's coastal patrol operation to avoid the civilian effort from wearing itself out. He further cited his opinion that a great injustice had been done to the coastal patrol pilots "in persuading them to go into the reserve [ACER] in the belief that this situation would exist for the duration of the emergency, and then to refuse to give them spare parts to keep their airplanes in commission. This results in these men being privates when they might have had a chance to be commissioned as pilots in the Air Transport Command or Flying Training Command."[64] Brig Gen Thomas J. Hanley, Deputy Chief of the Air Staff, met with Johnson to discuss the reasoning behind denial of the certificate of necessity, but Johnson stuck to his position for immediate dissolution of CAP.[65] Arnold in turn asked Maj Gen Westside T. Larson, AAFAC's commanding general, if CAP was performing a necessary mission for the Army and if CAP's antisubmarine activities were essential to combating submarines and thus continuing for the foreseeable future. Arnold also sought the recommendations of Fourth Air Force commander, Maj Gen Barney M. Giles, as to the retention of CAP in its present status or not.[66]

Figure 26. Stinson Voyager 10A owned by Bruce P. Ellen of Canton, North Carolina, seen at Coastal Patrol Base No. 21, Beaufort, North Carolina, in Navy three-tone camouflage, armed with a Navy Mk 15 100-pound practice bomb. (Photograph courtesy of the Charles Small Family, Richmond, Virginia, via the Morse Center.)

Ironically, Admiral Doenitz threw CAP a lifeline. Decreased enemy submarine activity in November gave the Army and Navy sufficient rationale to reduce CAP coastal patrol operations and deliberate its future.[67] To curtail unproductive patrols during this period of negligible enemy activity, AAFAC ordered CAP National Headquarters to reduce flying at all bases to conserve fuel, equipment, and personnel. AAFAC specifically cited the opportunity thus afforded "to perform necessary maintenance work and to bring the equipment to the highest practicable state of mechanical efficiency ready for more intensive operations when needed."[68]

Less than two weeks later, Vice Admiral Andrews and the Navy issued the same guidance for CAP coastal patrol bases in the Eastern Sea Frontier. Deeming the situation of reduced enemy activity advisable for decreased operations, Andrews issued instructions to limit routine patrols and fly only two, two-ship patrols daily, one each at dawn and dusk. Bases would conduct escort missions and special antisubmarine missions as directed and rescue or special search missions as requested. He also expressed a desire for the coastal patrol to direct its primary operations "to the areas approximately between the 5 fathom and 20 fathom curves," the body of water where enemy submarines would be most likely to operate with success, 15–20 miles distance from the coastline.[69]

The curtailments of Eastern Sea Frontier operations coincided with one substantial expansion of base operations in the Gulf of Mexico. The Army had completed arrangements with the Mexican government in November to permit CAP coastal patrol aircraft to patrol or escort vessels in Mexican waters or make emergency landings on Mexican soil.[70] By the end of that month, the Navy had aircraft from Coastal Patrol Base No. 12, Brownsville, Texas, flying special escort missions as far south as Tampico, Mexico. Within weeks Base 12's patrol area formally extended to latitude 22° north, over 250 miles south of the border. All aircraft on this new foreign coastal patrol required additional liability and property damage insurance coverage.[71] Escort duty of specific ships or small convoys continued for at least a month, with CAP personnel staying over at Tampico for refueling and crew rest.[72]

CAP National Headquarters ordered coastal patrol bases in the Eastern Sea Frontier to curtail all flying operations on 1 December and use the time to improve equipment and facilities while personnel trained extensively.[73] The next day, it issued guidance on daily line inspections for all aircraft, covering propellers, engines, landing gear, wings, fuselage and control surfaces, signal lights, armament, and emergency equipment.[74] AAFAC forwarded Andrews's memorandum to the Gulf Task Force to encourage a similar directive (thereafter issued in January 1943).[75]

The CAP coastal patrol's unremitting equipment issues, however, had shaped the fate of the organization and its coastal patrol mission. On 12 December, King directed the Gulf Task Force and Eastern Sea Frontier commanders to curtail CAP operations by at least 30 percent over the next three months and to reduce the armament furnished to the CAP to "a minimum consistent with actual operational requirements," although the CNO probably did not realize the armament came from the Army. Still recognizing the valuable contribution of the CAP coastal patrols, King justified his decision in view of prospective increases in military aircraft in the sea frontiers and the "serious difficulties to be expected in maintaining civilian aircraft."[76] Transmitting King's memorandum to the AAFAC, Rear Adm Patrick N. L. Bellinger explained it did not appear advisable to discontinue

CAP, but an increase in Navy aircraft deployed to the sea frontiers would permit the conclusion of CAP's operations by about 1 July 1943.[77]

CAP's shortages in the fall of 1942 paralleled those of the War Training Service (WTS, formerly the Civilian Pilot Training Program). Want of parts and airframes delayed training for Army Air Force enlisted reservists at WTS operations nationwide. To minimize nonessential flying and conserve aircraft, parts, and accessories, the War Production Board issued General Limitation Order L-262 on 26 January 1943. The order froze the sale or rental of aircraft of 500 horsepower or less and Link Trainer flight simulators and provided a mechanism for the CAA to purchase these aircraft to route to WTS operations. The order impacted CAP by keeping in the hands of present owners those aircraft needed for active duty missions and placing priority for parts and accessories on coastal patrol aircraft for the duration of the operation. All other aircraft serving with CAP became open for requisition by WTS.[78]

The concerns of late 1942 for the maintenance and condition of the airframes did not diminish. Also, a tragic trend emerged: after suffering only one fatality in July, nine CAP coastal patrol aircrew died in accidents and crashes from October to December.[79] In January 1943, Major Burnham at Base No. 4 at Parksley explained in detail why his unit could only provide four aircraft for patrol operations. Of 20 assigned aircraft, 15 were grounded for lack of parts, major overhauls and repairs, or changes in policy. Of the five available, three would soon require major overhauls. The 20 aircraft included seven different types of 12 models, including four types of engines and over 10 different power plant models. *Substitution with safety* became the base mantra for acquiring parts to keep aircraft aloft. "If a Base is to bear its responsibilities and is to be effective from a military point of view, a source of supply of aircraft material must be available," Burnham concluded.[80]

OCD began discussions with the War Department in January 1943, examining the relationship between CAP and its missions in support of the Army. Although aware of CAP's continuing struggles with aircraft maintenance, OCD lacked sufficient resources to help. As it stood, multiple voices within the War Department had advocated transferring CAP to the Army. Whether a matter of pure financial aid, procurement of parts, or even complete aircraft, all matters would be simplified if CAP were under military control. In mid-

January, Lovett, a supporter of CAP's coastal patrol, suggested to Landis that the Army Air Forces would like to take over CAP, providing a possible solution to CAP's operational struggles.[81] Before crafting a reply, Landis assembled a three-person committee to address the CAP-Army relationship regarding the aircraft maintenance, organizational efficiency, and overall continuation of CAP.[82] In a confidential letter to Gill Robb Wilson asking to help motivate the committee, Landis emphasized a need for swift action so he could reference its work in his reply to Lovett. "Otherwise," closed Landis, "there is a great likelihood that the Army will just ride roughshod over us."[83]

In his first reply to Lovett's suggestion of an Army transfer, Landis expressed that his dominating motive was promoting CAP's efficiency to support the prosecution of the war. Proud of CAP's operations and activities, the OCD director desired to see CAP remain true to its operations. Landis eloquently spoke of the "the soul of the Civil Air Patrol" as "the civilian incentive and ingenuity which has conquered obstacle after obstacle in the development of its organization and in the performance of its missions." Acknowledging how some of CAP's activities depended on not only the cooperation of the Army "but an appreciation by the Army of the functions that can be ably carried on by civilians," Landis reiterated his concerns about CAP's character. Speaking with a sense of responsibility for what CAP had achieved, he asked Lovett for "reasonable assurance . . . that these aspects of Civil Air Patrol will be fostered and cared for" by the Army.[84]

In his response, Lovett affirmed that he shared Landis's attitude for the transfer of CAP to the War Department. "The only question to consider," wrote Lovett, "is whether or not the change would be of material assistance in making effective use of the service which the Civil Air Patrol can render in the prosecution of the war." The Assistant Secretary for Air concluded the transfer would more effectively utilize CAP's services for the war effort. He articulated several ancillary reasons for the transfer, namely that the Army Air Force paid for 95 percent of CAP's operations, which were conducted under the direction of Army officers. A transfer to the War Department—with the complete accord of the Navy, he noted—would eliminate the unnecessary difficulty encountered in provisioning equipment and supplies; bringing CAP directly under War Department control would enable the purchase of materials or supplies from the department's depots. Regarding Landis's request for a reasonable assurance, Lovett wrote, "I can only answer this by saying that it is our intention to continue to make use of the CAP in every field where the expense in men, money, and materials is justified as a part of the over-all war effort, including in that objective the importance of increasing the flying experience of a large number of civilians and stimulating and developing interest in aviation among all our citizens, particularly the younger men."[85]

In reply, Landis raised no objection to Lovett's proposal, there being "only one answer that I can give." He thanked Lovett for his considered judgment of the Army's employment of CAP, because "I should hate to see this group of men become a mere stepchild of the Army."[86]

As these discussions about a CAP transfer to the Army unfolded, Andrews still sought to maximize CAP's utility. In the event of a submarine sighting, Andrews wanted every aircraft ready to attack; while no Army or CAP aircraft bomb drops were recorded for December 1942, attacks had resumed in January 1943.[87] In February 1943, he directed the Navy's Bureau of Ordnance to issue sufficient ordnance and flares to all CAP coastal patrol bases in the Eastern Sea Frontier for the first quarter of the year. As it stood in late February, Andrews could claim operational control over 328 aircraft from the Army, Coast Guard, and Navy—not counting CAP's force—as well as 200 assorted warships. An examination of CAP's flying hours shows a marked increase in flying in April and May, due in part to improved weather but without a marked increase in enemy submarine activity.[88] With this level of coverage, the coastal shipping lanes of the Eastern Sea Frontier meant death to any U-boat.[89]

After OCD and the Army worked out the assorted legislative matters with the Bureau of Budget, President Roosevelt issued Executive Order 9339 on 29 April, transferring CAP from OCD to the War Department.[90] On that day, CAP reported having 1,683 personnel and 423 aircraft assigned to coastal patrol duty.[91] In connection with the executive order, Lovett wrote Landis, acknowledging the transfer as "recognition of a job well done . . . motivated by a desire to make the Civil Air Patrol more directly available to perform its services to the armed forces."[92] In reply, Landis admitted, "I hate to say goodbye to the Civil Air Patrol purely from a personal standpoint as I have developed an affection for many of the men who sweated at it."[93] (This transfer's lasting importance was cemented just over two years later on 2 June 1945, when Pres. Harry S. Truman issued Executive Order 9562 terminating the Office of Civilian Defense effective 30 June, thereby making CAP the only element of OCD to survive World War II.[94])

As Landis desired, the transfer authorized the continuation of CAP's operational missions and the national staff remained intact.[95] On 4 May, Secretary of War Henry L. Stimson tasked Arnold with supervising and directing the operations of the Civil Air Patrol on his behalf.[96] The same day, the Navy requested CAP patrols out of Beaufort, North Carolina, to assist in locating survivors of the Panamanian motor tanker *Panam*, sunk by a single torpedo from *U-129* earlier in the morning darkness. The intrepid CAP planes managed to sight 40 survivors and help direct the subchaser USS *SC-664* to the lifeboats to rescue the torpedoed merchant mariners.[97] This successful demonstration of joint operations aside, the Army Air Forces replied to King's memorandum in mid-May that they had no objection to closing the CAP coastal patrol.[98] Two days after the Army's decision, CAP National Headquarters directed all CAP coastal patrol bases to avoid expending substantial amounts of money for permanent base improvements.[99]

Arnold saw to the publication of Army Air Forces Regulation 20-18 on 25 May, establishing CAP as "an exempted activity under the supervision of the Commanding General of the Army Air Forces" which placed administration and supply of CAP field activities under the direction of Headquarters Army Air Forces.[100] The day after CAP's transfer to the War Department, King directed the Eastern and Gulf Sea Frontier commanders to relieve and replace all CAP coastal patrol units with military personnel and aircraft by 31 August.[101]

More urgently for those operational activities of CAP, the availability of aircraft parts and maintenance for active duty operations also improved with the transfer. Army Air Forces Regulation 20-18 included a section tasking Headquarters Army Air Forces with the administration and supply of CAP field activities. On 3 June, the Supply Division of the Air Service Command reached an agreement with CAP for the Army to set up four centralized parts depots to provide parts, service Army Air Force–installed equipment, and furnish flying and safety equipment for the coastal and southern liaison patrols. Through this arrangement, CAP had a mechanism to maintain its aircraft and protect its aircrews. Unfortunately, the regulation for furnishing supplies and services to CAP was not published and did not take effect until after the cessation of coastal patrol operations.[102]

For the coastal patrol operation, however, improvements failed to materialize over the ensuing months. Johnson advised all base commanders on 19 June that if they had any doubt as to the airworthiness

Figure 27. Survivors of the torpedoed Panamanian motor tanker *Panam* are picked up by subchaser USS *SC-664* after being spotted by a CAP patrol from Coastal Patrol Base No. 21, Beaufort, North Carolina, 4 May 1943. A CAP aircraft can be seen circling the subchaser, photographed from above by a Navy blimp. (Photograph courtesy of Naval History and Heritage Command.)

of a plane to ground it: "This may mean that important missions will not be flown, but this is as it must be."[103] In a candid letter of 28 June to Navy Lt R. E. Schreder, Base No. 4's Major Burnham again spoke frankly of the state of his unit's assigned aircraft. He considered it obvious that all CAP coastal patrol bases "are approaching a point where operations will have to be discontinued unless aircraft are provided." He detailed the difficulty in sourcing parts for civilian planes, with the only alternative being a dangerous policy to improvise or substitute parts. With spare engines "practically impossible" for bases to acquire, noted Burnham, "it is my opinion that CAP Coastal Patrol Bases will be unable to carry on this winter without new aircraft." He continued, "The fact that to the writer's knowledge 10 aircraft have been lost in Civil Air Patrol Bases at Rehoboth, Parksley, Manteo, and Beaufort in the past 30 days is an indication of the fact that the aircraft are just too tired to continue much longer. Every one of these aircraft crashed as the result of engine failure. In one of the accidents, 2 persons were killed, and all of the rest of the crews were saved."[104] Even with improved survival equipment, no one wished to repeat the winter of 1942 when 10 volunteers gave their lives to the country, with the bodies of four of them never recovered.[105] Clearly, it was better to be prudent than merely courageous.

In the summer of 1943, the Army and Navy reached a final agreement about aerial operations in antisubmarine warfare that profoundly affected CAP's coastal patrol operations. While AAFAC had just started to demonstrate its power and capability, the Navy's own air arm had grown strong enough to also conduct offensive operations with long-range, land-based patrol bombers. The duplication of effort between services for land-based antisubmarine warfare aircraft proved needlessly wasteful of resources and militarily inefficient. These facts, if known, Marshall noted in a memorandum to King, "would inevitably meet with public condemnation."[106] After months of heated discussions, the War and Navy Departments finally accepted an agreement on 9 July. While the Army agreed that aerial antisubmarine operations would remain the primary responsibility of the Navy, the latter agreed to the Army's authority to provide for long-range striking forces (strategic bombing) for defense of the Western Hemisphere. The Army would withdraw from antisubmarine air operations at such time as the Navy could take over such duties completely. The Army would transfer and exchange 77 specialized AAFAC B-24 heavy bombers in exchange for an equal number of combat-equipped B-24s allocated to the Navy. The services would complete the handover by October.[107]

Less than a week after the Army agreed to turn antisubmarine operations over to the Navy, CAP's coastal patrol personnel received new orders. On 15 July, CAP National Headquarters informed all bases of the Navy's decision to cease all patrol operations at sundown, 31 August.[108] Until then, all activities would continue at their present scale.[109] In mid-August, 12 base commanders received guidance on procedures for the liquidation of their bases and reassignment of personnel and aircraft to home squadrons or new active duty missions.[110] As the sun set on CAP's coastal patrols, the Eastern Defense Command and Army Air Forces redesignated AAFAC as I Bomber Command with orders to disband the 25th and 26th Antisubmarine wings.[111]

At the conclusion of its operations, CAP National Headquarters provided operational statistics to the Army and Navy. From March 1942 to August 1943, civilian volunteers flew 86,685 missions totaling 244,600 hours, at a loss of 90 aircraft with 26 personnel killed (see appendix B) and seven seriously injured. CAP aircrews reported sighting 91 vessels in distress, 173 suspected submarines, 363 survivors of attacks, 36 dead bodies, and 17 floating mines. At the request of the Navy, CAP conducted 5,684 special convoy escort missions.[112] Most notably, from the period of 1 October 1942 to 31 August 1943, when U-boats were operating in the North Atlantic, AAFAC recorded 375,269 hours of antisubmarine operational flying; CAP accounted for 196,636 hours of this total, including 41,897 hours of escort, 89,504 hours of patrols, and 60,548 hours of reconnaissance.[113]

For the men and women of the coastal patrol, the end of coastal patrol operations came with varying measures of acknowledgement and revelry. Several of the bases enjoyed a final banquet, some with live music and dancing.[114] At Base No. 3, Sen. Claude Pepper (D-FL) addressed the assembled base personnel prior to the last retreat. "You have carried out a dangerous and difficult task that demanded true courage and devotion to your country. You have helped save this nation from the attacks of an invader and helped lead it back toward the happy days of peace."[115] In Manteo, North Carolina, the local paper published an editorial on the closure of Base No. 16, acknowledging the relationship between the civilian volunteers and the local residents for whom they served:

> The many pleasant social relations that have been established between the members of the Civil Air Patrol and the citizens of this community have created warm friendships and the departure of this group is like unto the "farewells" that come when old friends have ended a visit. The citizens of this community wish them "Godspeed" and keep in mind the old Biblical quotation, "Well done, good and faithful."[116]

Fittingly, visitors to the contemporary Dare County Regional Airport will be greeted outside by a prominent marker listing the names of every Base No. 16 member, forever friends of Manteo, North Carolina.

Senior military officials in Washington also gave thanks to CAP's coastal patrol at the conclusion its operations. Andrews expressed his appreciation to the base commanders, recognizing how each unit "rendered invaluable services" to the Eastern Sea Frontier, and in the performance of the unit taskings, members "displayed a skill, energy, resourcefulness, and disregard for danger which are in the highest tradition of the American armed forces."[117] King, once skeptical of the civilian effort, commended the work of the CAP coastal patrol on 11 August 1943, expressing a "'WELL DONE' for their enthusiastic, loyal, and constant cooperation in combating the submarine menace, patrolling our coastline and assisting in the locating of survivors and ships in distress."[118] For King, never one to offer accolades except when appropriate, his praise represented his highest compliments.

Arnold, one of CAP's biggest proponents, offered perhaps the most poignant words on the value and accomplishments of the CAP coastal patrol in a speech of 16 December 1944. Contextualizing the civilian effort in relation to the war, the general remarked: "The Civil Air Patrol itself grew . . . out of the urgency of the situation . . . [and was] set up and went into operation almost overnight. It patrolled our shores and performed its anti-submarine work at a time of almost desperate national crisis. If it had done nothing beyond this, the Civil Air Patrol would have earned an honorable place in the history of American air power."[119] A gallant salute to the nation's civilian aviators who met the challenges posed by the War Department, answered the call to serve, and did not waver in the face of adversity.

Figure 28. The flag at Coastal Patrol Base No. 4, Parksley, Virginia, being lowered at the conclusion of the coastal patrol operation, 31 August 1943. (Photograph courtesy of William G. Bell via the Morse Center.)

Notes

Epigraph. E. H. Johnson, "Civil Air Patrol Operations in Fifth Naval District," 13 September 1944, Reel A4057, AFHRA.

1. In August 1942, CAP queried all CAP members with registered aircraft to locate more aircraft for coastal patrol and other active duty missions. OCD, CAPNHQ, Earle L. Johnson, to CAP Members Owning Registered Aircraft, memorandum, subject: Availability of Aircraft for Active Duty, 28 August 1942, Folder "Civil Air Patrol 1 August Thru," Box 21, Entry 16A, General Correspondence, 1941–May 1945, Civil Air Patrol, RG171, NARA.

2. Seven of the 21 coastal patrol bases relocated during their operational existence, all in 1942. See appendix A. CAPNHQ, untitled document marked "Confidential" listing movement of Coastal Patrol Bases, 15 February 1943, Reel 38920, AFHRA.

3. Civil Aeronautics Board, George M. Keightley, Investigator's Report, Civil Coastal Patrol, Panama City, Florida, Grand Isle, Louisiana, 1 September 1942 and Pascagoula, Mississippi, 31 August 1942, BLS; and Speiser, "'Joe' – Submarine Hunter," 8, CAP-NAHC.

4. Green, "The History of Coastal Patrol Base #14," 6–7; and Civil Aeronautics Board, George M. Keightley, Investigator's Report, Civil Coastal Patrol, Panama City, Florida, Grand Isle, Louisiana, 1 September 1942 and Pascagoula, Mississippi, 31 August 1942, BLS, CAP-NAHC.

5. Civil Aeronautics Board, Robert H. Peters, Inspection of North Carolina Wing, Civil Air Patrol Base at Manteo, North Carolina, 4 September 1942, BLS, CAP-NAHC.

6. E. H. Johnson, "Civil Air Patrol Operations in Fifth Naval District," 13 September 1944, Reel A4057, AFHRA; Bridges, interview; Wescott, interview; Fields, interview; and Keefer, *From Maine to Mexico*, 383, 387.

7. Blazich, "North Carolina's Flying Volunteers," 421–22; E. H. Johnson, "Civil Air Patrol Operations in Fifth Naval District," 13 September 1944, Reel A4057, AFHRA; and Warner and Grove, *Base Twenty-one*, 54–93.

8. Claire H. Hamlin, "The Civil Air Patrol at Coastal Patrol Base #20," KKH, CAP-NAHC; Orrin B. Maxwell to Earle L. Johnson, memorandum, subject: Expenditures in Establishing CCP Base #20, 17 December 1942; and Orrin B. Maxwell to Earle L. Johnson, memorandum, subject: Detailed Report on Destruction Caused by Fire on 10 December 1942, 14 December 1942, Reel 38920, AFHRA. Two months after the fire in the fading light of 12 February, dusk patrols from Base No. 20 spotted Naval Reserve Lt John Shelley alive but in shock after the crash of his aircraft in Blue Hills Bay. The CAP aircraft dropped flares to help guide a boat manned by townspeople from Surry, Maine, to Shelley's location, and they plucked the aviator from the icy waters, saving his life. Mellor, *Sank Same*, 82–83; Keefer, *From Maine to Mexico*, 483–84; "Navy Man Feared Lost After Crash in Blue Hill Bay," *Bangor (ME) Daily News*, 15 February 1943, 1; and "Two Men Escape Death as Navy Planes Collide," *Knoxville (TN) Journal*, 14 February 1943, 7.

9. OCD, CAPNHQ, Harry H. Blee to All Regional, Wing, and Base Commanders, memorandum, subject: Confidential Letter of Instruction no. 5, Organization and Operation of Coastal Patrol Bases, 11 May 1942, Reel 38907, AFHRA.

10. OCD, CAPNHQ, Operations Memorandum no. 7, Recruiting Civilian Pilot and Mechanic Personnel, 8 October 1942, Reel 38909, AFHRA.

11. Jividen, interview, 4–6; Speiser, "'Joe'–Submarine Hunter," 5–6; and Burnham, "History of CAP Coastal Patrol No. 4" (ca. 1943), 8–9.

12. Civil Aeronautics Board, Perry Hodgden, Inspection of Civil Air Patrol Maintenance at San Antonio, Corpus Christi, and Brownsville, Texas, 26 August 1942, BLS, CAP-NAHC.

13. Earle L. Johnson to All CAP Coastal Patrol and Liaison Patrol Commanders, memorandum, subject: Finances of CAP Coastal Patrols and Liaison Patrols, 20 January 1943, BLS, CAP-NAHC.

14. Burnham, "History of CAP Coastal Patrol No. 4," 11–15; and Nanney, *CAP Coastal Patrol Base No. 5*, 17.

15. Civil Aeronautics Board, George M. Keightley, Inspections of Civil Coastal Patrol Bases at North Miami, Lantana, Daytona Beach, Tampa, Florida and Brunswick, Georgia, 29 August 1942, BLS, CAP-NAHC.

16. Speiser, "'Joe'–Submarine Hunter," 8; and Mellor, *Sank Same*, 82.

17. Boudreau, *CAPCP Base-17*, 29.

18. Mellor, *Sank Same*, 66–68; Keefer, *From Maine to Mexico*, 40, 58; and Robert B. Neprud, "Rehoboth…re. Capt. Everett M. Smith (from *Sank Same*)," Reel 38910, AFHRA.

19. Robert B. Neprud, "Rehoboth Fact Sheet," Reel 38910, AFHRA. In a report of 1 July 1943 from CAP National Headquarters, Base No. 2 lists 12 out of its 18 assigned aircraft as the Fairchild 24. Of similar nature, Coastal Patrol Base No. 8 at Charleston, SC, lists 19 aircraft, 16 of which are the Stinson 10A. Harry H. Blee to Howard S. Sterne, memorandum, subject: Aircraft at Coastal and Liaison Patrol Bases, 1 July 1943, with attached list, "Aircraft at CAP Coastal and Liaison Patrol Bases," 1 July 1943, BLS, CAP-NAHC.

20. Civil Aeronautics Board, George M. Keightley, Investigator's Report, Civil Coastal Patrol No. 9, James Island, Charleston, South Carolina," 19 August 1942, BLS, CAP-NAHC.

21. Civil Aeronautics Board, Julian R. Wagy, Investigator's Report, Civil Air Patrol Base Beaufort, North Carolina, 27 February 1943, BLS, CAP-NAHC.

22. Civil Aeronautics Board, Fred G. Powell, Inspection of Coastal Patrol of Civil Air Patrol at Atlantic City, New Jersey, 5 August 1942; Civil Aeronautics Board, George M. Keightley, Inspections of Civil Coastal Patrol Bases at North Miami, Lantana, Daytona Beach, Tampa, Florida and Brunswick, Georgia, 29 August 1942; Civil Aeronautics Board, George M. Keightley, Inspection of CAP Coastal Patrol No. 8, Charleston, South Carolina, 19 August 1942; Civil Aeronautics Board, Safety Bureau, Investigator in Charge, Fourth Branch Office to Chief, Investigation Section, memorandum, subject: Civil Air Patrol–Maintenance, 31 July 1942; and Civil Aeronautics Board, George M. Keightley, Investigator's Report, Civil Coastal Patrol, Panama City, Florida, Grand Isle, Louisiana, 1 September 1942 and Pascagoula, Mississippi, 31 August 1942, BLS, CAP-NAHC.

23. Civil Aeronautics Board, Safety Bureau, Investigator in Charge, Fourth Branch Office to Chief, Investigation Section, memorandum, subject: Civil Air Patrol Maintenance at San Antonio, Corpus Christi and Brownsville, Texas, 26 August 1942, BLS, CAP-NAHC.

24. OCD, CAPNHQ, Civil Air Patrol Coastal Patrol Operations, 20 October 1942, Reel 38919, AFHRA.

25. OCD, CAPNHQ, Operations Directive no. 23A, CAP Coastal Patrols, 26 August 1942; OCD, CAPNHQ, Operations Directive no. 23A, Change 1, CAP Coastal Patrols, 26 February 1942, AWS, CAP-NAHC; and OCD, CAPNHQ, Operations Directive no. 23, Change no. 3, Task Forces on Coastal Patrol Duty, 24 August 1942,

Binder "54 Civil Air Patrol," Box 1, Entry 54: Processed Documents Issued Serially by the Civil Air Patrol, Office of the Director, RG171, NARA.

26. OCD, CAPNHQ, Operations Directive no. 33, Maintenance of Physical Conditions of Coastal Patrol and Liaison Patrol Flight Personnel, 11 November 1942, Binder "54 Civil Air Patrol," Box 1, Entry 54, Office of the Director, Processed Documents Issued Serially by the Civil Air Patrol, RG171, NARA.

27. OCD, CAPNHQ, Operations Directive no. 29, Check List for Pilots, 1 October 1942, Binder "54 Civil Air Patrol," Box 1, Entry 54, Office of the Director, Processed Documents Issued Serially by the Civil Air Patrol, RG171, NARA.

28. Floyd O. Johnson to Jack Vilas, 15 July 1942, BLS, CAP-NAHC.

29. George M. Keightley, "Investigator's Report Civil Coastal Patrol, Panama City, Fla.," 1 September 1942, BLS, CAP-NAHC.

30. W. E. Koneczny to W. F. Andrews, memorandum, subject: Bomb attachment on CAP Airplanes, 14 August 1942, BLS, CAP-NAHC.

31. OCD, CAPNHQ, Operations Directive no. 29, Check List for Pilots, 1 October 1942, Binder "54 Civil Air Patrol," Box 1, Entry 54, Office of the Director, Processed Documents Issued Serially by the Civil Air Patrol, RG171, NARA.

32. Royal E. Ingersoll to Atlantic Fleet, memorandum, subject: Civilian Air Patrol Coastal Patrol, 28 October 1942, file "COMINCH 1942 Secret A16-1 National Defense," Box 255, Headquarters COMINCH, 1942–Secret, A14-1 to A16-1(1), RG38, NARA. Aircraft had to drop ordnance at sufficient altitude to ensure the bomb armed and to avoid damage from the resulting explosion. The actual safe altitude varies depending on the source. AAFAC instructed CAP aircraft when carrying demolition bombs to patrol at an altitude of no less than 600 feet. G. A. McHenry, Headquarters, Army Air Force Antisubmarine Command, to Commanding Officer, 25th Antisubmarine Wing, Commanding Officer, 26th Antisubmarine Wing, and All CAPCP Units, memorandum, subject: Letter of Instructions No. 1C, 31 December 1942, Reel A4063; and Coastal Patrol No. 18, memorandum no. 12-01, subject: Directive, 7 September 1942, Reel 38920, AFHRA; Mellor, *Sank Same*, 108–9; and Milton P. Arnette, "Bombing Notes," Binder "Civil Air Patrol Coastal Patrol Base 16 History, Volume 2–Various Documents, Orders, and Information, 1942–1943," Dare County Regional Airport, Manteo, NC.

33. Schoenfeld, *Stalking the U-Boat*, 7, 14–15.

34. As of 1 October 1942, only 193 of 415 CAP aircraft assigned for coastal patrol duty were armed. T. Jefferson Newbold to Harry H. Blee, memorandum, subject: Bomb Rack Installation, 16 October 1942, Reel A4064, AFHRA.

35. Army Air Forces Antisubmarine Command, February 1943 Monthly Intelligence Report, 24–25, Reel A4057, AFHRA.

36. "Inexpensive Bombing for Civil Air Patrol," *Air Force: Official Service Journal of the U.S. Army Air Forces* (January 1943): 34; and "Bomb Sight for Light Airplanes," *Aero Digest* 41, no. 2 (August 1942): 214. CAP National Headquarters requested data about installed bomb racks, bombsights, and ordnance stocks from all 21 coastal patrol bases in October 1942. OCD, CAPNHQ, Coastal Patrol Circular no. 17, Report of Special Equipment and Supplies, 29 October 1942, AWS, CAP-NAHC; and summation of CAP coastal patrol base reports, Folder "Civil Air Patrol Coastal and Liaison Patrol Special Equipment, 1942/11, CAP Item Number 128," Reel 44599, AFHRA.

37. Keefer, *From Maine to Mexico*, 122–23, 147–48, 180, 264; Mosley, *Brave Coward Zack*, 55–56; Mellor, *Sank Same*, 106–9; Myers, interview; Parkinson, interview; and Gantt, interview.

38. OCD, CAPNHQ, Civil Air Patrol Coastal Patrol Operations, 20 October 1942, Reel 38919, AFHRA.
39. Frederick Lyall to Earle L. Johnson, 27 October 1942, BLS, CAP-NAHC.
40. Jacob Pypelink to Earle L. Johnson, 13 November 1942, BLS, CAP-NAHC.
41. "What Is a 'Barracuda Bucket'"; and Keil, "Civil Air Patrol Coastal Patrol No. Three."
42. E. H. Johnson, "Civil Air Patrol Operations in Fifth Naval District," 13 September 1944, Reel A4057, AFHRA.
43. E. C. Foster, Headquarters, 25th Antisubmarine Wing, to Westside T. Larson, memorandum, subject: Safety Equipment for Civil Air Patrol Coastal Patrol Units, 27 December 1942, Reel A4064; Haddaway, interview; Keefer, *From Maine to Mexico*, 178, 185–86, 246; Blazich, "North Carolina's Flying Volunteers," 433–35; and "Lack of Rafts Blamed for CAP Pilots' Deaths," *Boston Globe*, 7 February 1943, C2. Frank M. Cook, Julian L. Cooper, Drew L. King, Alfred H. Koym, Clarence L. Rawls, Guy T. Cherry Jr., and James C. Taylor all crash-landed and managed to exit their aircraft and inflate their life vest or barracuda bag. All died before rescue crews could pull them from the water.
44. OCD, CAPNHQ, Pneumatic Life Rafts, 15 December 1942, Reel 44599, AFHRA.
45. OCD, CAPNHQ, Harry H. Blee, Coastal Patrol Circular no. 22, Swimming Instructions for CAP Coastal Patrol Flight Personnel, 28 November 1942, Binder "54 Civil Air Patrol," Box 1, Entry 54, Office of the Director, Processed Documents Issued Serially by the Civil Air Patrol, RG171, NARA.
46. Fred G. Powell, "Inspection of Civil Air Patrol Base No. 7, Chapman Field, Perrine, Florida," attachment to W. K. Andrews to Harry H. Blee, 8 March 1943, BLS, CAP-NAHC.
47. OCD, CAPNHQ, Harry H. Blee to All CAP Coastal Patrol Commanders, memorandum, subject: Pneumatic Life Rafts – CAP Coastal Patrols, 30 March 1943; and Howard S. Stern to Chief of the Bureau of Aeronautics, Navy Department, memorandum, subject: Rescue of CAP Air Crews, 31 May 1943, BLS, CAP-NAHC.
48. Blazich, "North Carolina's Flying Volunteers," 434–35.
49. *19th Patrol Force*, 33.
50. OCD, *CAP Bulletin* 2, no. 15, 9 April 1943, Folder 6, Box 6, ELJ, WRHS; and Hopper, *Civil Air Patrol Historical Monograph Number 1*, 5–12.
51. Earle L. Johnson to All Task Force Base Commanders, memorandum, subject: Draft Deferment, 18 July 1942, Reel 38919, AFHRA.
52. Earle L. Johnson to Henry L. Stimson, 28 July 1942, BLS, CAP-NAHC.
53. Donald H. Connolly to Earle L. Johnson, 21 August 1942; and OCD, CAP-NH, Harry H. Blee to All CAP Coastal Patrol Commanders, memorandum, subject: Enlistment of Civil Air Patrol Pilots and Mechanics in the Air Corps Enlisted Reserve, 27 August 1942, BLS, CAP-NAHC.
54. OCD, CAPNHQ, Harry H. Blee to All CAP Coastal Patrol and Liaison Patrol Commanders, memorandum, subject: Enrollment of Coastal Patrol and Liaison Patrol Pilots and Mechanics in Air Corps Enlisted Reserve, 19 November 1942, BLS, CAP-NAHC.
55. OCD, CAPNHQ, Harry H. Blee to All CAP Coastal Patrol and Liaison Patrol Commanders, memorandum, subject: Enrollment of Coastal Patrol and Liaison Patrol Pilots and Mechanics in Air Corps Enlisted Reserve, 19 November 1942; teletype from Harry H. Blee to All Coastal Patrol Base Commanders, 8 December 1942; Earle L. Johnson to J.N. Belknap, memorandum, subject: Use of Air Corps Enlisted Reserve, 8 July 1943, BLS, CAP-NAHC.

56. H. B. Sepulveda to Henry H. Arnold, memorandum, subject: Status of Individual Members of the Civil Air Patrol, 6 July 1942; Randolph Williams to Follett Bradley, memorandum, subject: 1st Indorsement, 25 July 1942, Reel A4064, AFHRA.
57. OCD, CAPNHQ, Operations Orders no. 1, Activation of CAP Coastal Patrols, 30 November 1942, Folder 2, Box 6, ELJ, WRHS.
58. Floyd O. Johnson to Earle L. Johnson, 13 October 1942, BLS, CAP-NAHC.
59. Harry H. Blee to Floyd O. Johnson, memorandum, subject: Repair and Replacement Parts–Stinson "Voyagers," 24 October 1942, BLS, CAP-NAHC.
60. Eugene L. Eubank to Westside T. Larson, memorandum, subject: Civil Air Patrol, 26 November 1942, Reel A4064, AFHRA. According to an entry of 23 October 1942, Col Horace W. Shelmire, Arnold's executive assistant, called CAP National Headquarters and reported that Arnold signed the certificate. Record of telephone calls, October 1942, Folder 6, Box 1, ELJ, WRHS.
61. Eugene L. Eubank to Westside T. Larson, memorandum, subject: Civil Air Patrol, 26 November 1942, Reel A4064, AFHRA.
62. Irving H. Taylor to Earle L. Johnson, 7 November 1942, BLS, CAP-NAHC.
63. E. H. Johnson, "Civil Air Patrol Coastal Patrol in Fifth Naval District," 13 September 1943, Reel A4057, AFHRA.
64. Internal summary report from Thomas J. Hanley Jr. for Henry H. Arnold, 10 November 1942, Folder "SAS 324.3–Civil Air Patrol," Box 93, HHA, LOC.
65. Internal summary report from Thomas J. Hanley, Jr. for Henry H. Arnold, 10 November 1942, Folder "SAS 324.3–Civil Air Patrol," Box 93, HHA, LOC; and entry for 10 November 1942, record of telephone calls, November 1942, Folder 6, Box 1, ELJ, WRHS.
66. Robert W. Harper to Westside T. Larson, memorandum, subject: Utilization of the Civil Air Patrol, 14 November 1942, Reel A4064, AFHRA.
67. The Army Air Forces Antisubmarine Command monthly report for November 1942 reports how "the withdrawal of enemy submarines from the Atlantic Coast of the United States became virtually complete during the month." Army Air Forces Antisubmarine Command, "Monthly Intelligence Summary, November 1942," 5, Reel A4048, AFRHA.
68. Westside T. Larson to Earle L. Johnson, memorandum, subject: Reduction of Flight Operations and Flying Hours, CAPCP Units, 4 November 1942, Reel A4064, AFHRA.
69. Adolphus Andrews to Westside T. Larson, memorandum, subject: Reduction of Flight Operations and Flying Hours, Civil Air Patrol Coastal Patrol Units of Eastern Sea Frontier, 16 November 1942, Folder "SAS 324.3–Civil Air Patrol," Box 93, HHA, LOC.
70. John A. Feagin to Westside T. Larson, memorandum, subject: CAP-CP Missions in Mexican Waters, 11 November 1942; U.S. Naval Attaché, Mexico City, Mexico to Commander, Gulf Sea Frontier, Miami, Florida, memorandum, subject: Enclosure–Translation of Memorandum from Mexican Foreign Minister to the American Ambassador dated 21 October 1942, 30 October 1942, Reel A4064, AFHRA.
71. Commander Gulf Sea Frontier to American Legation United States Naval Attaché Mexico City, priority teletype, December 1942, Reel A4064, AFHRA; OCD, CAPNHQ, Summary–Operations Conducted by Civil Air Patrol for Army Agencies, 5 May 1943, BLS, CAP-NAHC.
72. Dan C. Putnam, "My Reminiscences of CAP Base 12," January 1986, CAP-NAHC; Keefer, *From Maine to Mexico*, 282, 284–85; Myers, interview, 5–6.

73. A handful of bases did not curtail flying and exceeded the permitted maximum of eight daily patrols until late December 1942. Harry H. Blee to Commanding Officers, CAP Coastal Patrols Operating in Eastern Sea Frontier Area, memorandum, subject: Reduction of Flight Operations, 26 December 1942, BLS, CAP-NAHC.

74. OCD, CAPNHQ, Harry H. Blee, Coastal Patrol Circular no. 21, Reduction of Flight Operations, CAP Coastal Patrols Operating in Eastern Sea Frontier Area with attachments A and B, 24 November 1942, AWS, CAP-NAHC; and OCD, CAPNHQ, Operations Directive no. 35, Daily Line Inspection of Aircraft CAP Operating Bases and Stations, Binder "54 Civil Air Patrol," Box 1, Entry 54, Office of the Director, Processes Documents Issued Serially by the Civil Air Patrol, RG171, NARA.

75. G. A. McHenry to Commanding Officer, Gulf Task Force, memorandum, subject: Reduction of Flight Operations, CAP CP Units Operating in Eastern Sea Frontier Area, 18 November 1942; Commander, Gulf Sea Frontier to Commanding Officer, 26th Antisubmarine Wing, AAFAC, memorandum, subject: Civil Air Patrol–Reduction in Flight Schedules, 8 January 1943, Reel A4064, AFHRA.

76. Ernest J. King to Adolphus Andrews and James L. Kauffman, memorandum, subject: Civilian Air Patrol, 12 December 1942, file "COMINCH 1942 Secret A16-1 National Defense," Box 255, Headquarters COMINCH, 1942–Secret, A14-1 to A16-1(1), RG38, NARA.

77. Patrick N. L. Bellinger to Army Air Force Liaison Officer, memorandum, subject: Operation of Civil Air Patrol in Sea Frontiers, 12 December 1942, file "COMINCH 1942 Secret A16-1 National Defense," Box 255, Headquarters COMINCH, 1942–Secret, A14-1 to A16-1(1), RG38, NARA. According to a Bureau of the Budget memorandum, King's directives to reduce CAP operations in thirds would completely remove CAP from coastal patrol work by October 1943. E. J. Donnelly to Herman Kehrli, memorandum, subject: Civil Air Patrol, 4 February 1943, Folder "OCD Civil Air Patrol," Box 10, Entry 107A, Budgetary Administration Records for Emergency and War Agencies and Defense Activities, 1939–1949 (39.19), RG51, NARA.

78. Pisano, *To Fill the Skies with Pilots*, 106–8; Earle L. Johnson to All Unit Commanders; All Base and Station Commanders, memorandum, subject: Aircraft Limitation Order (GM-72), 1 February 1943; Earle L. Johnson to All Unit, Base, and Station Commanders, memorandum, subject: Registration and Rental of CAP Planes (GM-74), 8 February 1943, Binder "Civil Air Patrol–Establishment of Charts, Staff, General Memoranda, Training Memoranda," Box 2, Entry 205, Processed Documents Issued by the OCD–CAP, RG171, NARA; OCD, *CAP Bulletin* 2, no. 6, 5 February 1943; and OCD, *CAP Bulletin* 2, no. 13, 26 March 1943, Folder 6, Box 6, ELJ, WRHS.

79. OCD, Civil Air Patrol Monthly Report, November 1942 (Restricted–Not for Publication), Folder "507 Civil Air Patrol," Box 70, Entry 51, General Correspondence, Director's Office, Feb. 1942–June 1944 500-520, RG171, NARA; Neprud, *Flying Minute Men*, 120; and Hudson, *Civil Air Patrol Fatalities*, 3. In 1943, CAP suffered 16 additional coastal patrol fatalities, 11 in the winter months of January through March when aircraft maintenance remained tenuous.

80. Isaac W. Burnham II to Walter W. Burbank, 16 January 1943, Binder "CAPCP Volume 1 Index Items 1-8," CHF, CAP-NAHC.

81. Earle L. Johnson to John F. Curry, 26 January 1943, Folder 2, Box 2, ELJ, WRHS; James M. Landis to William B. Stout, 15 January 1943; James M. Landis to Gill Robb Wilson, 15 January 1943; James Landis to Robert A. Lovett, 22 January 1943, Folder "507 Civil Air Patrol," Box 70, Entry 51, National Headquarters, General Correspondence, Director's Office, Feb. 1942 – June 1945 500-520, RG171; and

Ivan Hinderaker to Herman Kehrli, memorandum, subject: Transfer of Civil Air Patrol from OCD to War Department, 28 January 1942, Folder "OCD Civil Air Patrol," Box 10, Entry 107A, RG51, NARA.

82. Earle L. Johnson to John F. Curry, 26 January 1943, Folder 2, Box 2, ELJ, WRHS; James M. Landis to William B. Stout, 15 January 1943; James M. Landis to Gill Robb Wilson, 15 January 1943; James Landis to Robert A. Lovett, 22 January 1943, Folder "507 Civil Air Patrol," Box 70, Entry 51, National Headquarters, General Correspondence, Director's Office, Feb. 1942 – June 1945 500-520, RG171; Ivan Hinderaker to Herman Kehrli, memorandum, subject: Transfer of Civil Air Patrol from OCD to War Department, 28 January 1942, Folder "OCD Civil Air Patrol," Box 10, Entry 107A, RG51, NARA. Landis's CAP committee consisted of William B. Stout, Harry K. Coffey, and Earl Findley.

83. James M. Landis to Gill Robb Wilson, 15 January 1943, Folder "507 Civil Air Patrol," Box 70, Entry 51, National Headquarters, General Correspondence, Director's Office, Feb. 1942–June 1945 500-520, RG171, NARA.

84. James Landis to Robert A. Lovett, 22 January 1943, Folder "507 Civil Air Patrol," Box 70, Entry 51, National Headquarters, General Correspondence, Director's Office, Feb. 1942–June 1945 500-520, RG171, NARA.

85. Robert A. Lovett to James Landis, 26 January 1943, Folder "507 Civil Air Patrol," Box 70, Entry 51, National Headquarters, General Correspondence, Director's Office, Feb. 1942–June 1945 500-520, RG171, NARA.

86. James Landis to Robert A. Lovett, 28 January 1943, Folder "507 Civil Air Patrol," Box 70, Entry 51, National Headquarters, General Correspondence, Director's Office, Feb. 1942–June 1945 500-520, RG171, NARA.

87. Adolphus Andrews to William H. P. Blandy, memorandum, subject: Ordnance Equipment, requirements of Civil Air Patrol Coastal Patrol during First Quarter 1943, 28 December 1942; Adolphus Andrews to distribution list, memorandum, subject: Ordnance Equipment, Requirements of Civil Air Patrol, 14 February 1943; Reel A4064; Army Air Forces Antisubmarine Command, Monthly Summary, December 1942, 3; Army Air Forces Antisubmarine Command, Monthly Summary, January 1943, 3, Reel A4048, AFHRA; OCD, CAPNHQ, Harry H. Blee to Commanding Officers, CAP Coastal Patrols Nos. 1, 2, 4, 6, 8, 16, 17, 18, 19, 20, 21, memorandum, subject: Ordnance Equipment and Service, CAP Coastal Patrols Operating Within Eastern Sea Frontier, 25 February 1943, BLS, CAP-NAHC.

88. A comparison of reported CAP coastal patrol escort and reconnaissance patrol hours in the Army Air Forces Antisubmarine Command monthly report from October 1942 to August 1943 shows a marked decline in the number of patrol hours from November 1942 to March 1943, at which point the hours climb again in May, decline in June, rise again in July, and then decline. Overall, cumulative totals are never as high as they were for November 1942. Army Air Forces AntiSubmarine Command, "Monthly Summary," October 1942 through August 1943, Reel A4048, AFHRA.

89. Blair, Hitler's *U-Boat War: The Hunted, 1942-45*, 245–46. All subsequent references to this volume appear as Blair, *The Hunted*.

90. Herman Kehrli to William F. McCandless, memorandum, subject: Proposed Transfer of the Civil Air Patrol from the Office of Civilian Defense to the Army Air Forces, 30 January 1943, Folder "OCD Civil Air Patrol," Box 10, Entry 107A, Budgetary Administration Records for Emergency and War Agencies and Defense Activities, 1939-1949 (39.19), RG51; and Franklin D. Roosevelt, Executive Order 9339, "Transfer of Civil Air Patrol from the Office of Civilian Defense to the Department of War," *Federal Register* 8, no. 86 (1 May 1943): 5659.

91. Harry H. Blee to Earle L. Johnson, memorandum, subject: Coastal Patrol Activities of Civil Air Patrol, 22 to 28 April 1943, inclusive, 29 April 1943, Folder "Civil Air Patrol," Box 7, Entry 233, Director's Office, Feb. 1942–June 1944, CAP–Budget Estimates, RG171, NARA.

92. Robert A. Lovett to James M. Landis, 29 April 1943, Folder "507 Civil Air Patrol," Box 70, Entry 51, National Headquarters, General Correspondence, Director's Office, Feb. 1942–June 1945 500-520, RG171, NARA.

93. James M. Landis to Robert A. Lovett, 30 April 1943, Folder "507 Civil Air Patrol," Box 70, Entry 51, National Headquarters, General Correspondence, Director's Office, Feb. 1942–June 1945 500-520, RG171, NARA.

94. Harry S. Truman, Executive Order 9562, "Termination of the Office of Civilian Defense," *Federal Register* 10, no. 112 (6 June 1945): 6639.

95. David L. Robinson to Fred Levi, memorandum, subject: Transfer of CAP from OCD to War Department, 19 March 1943; Wayne Coy to Franklin D. Roosevelt, memorandum, subject: Transfer of the Civil Air Patrol to the War Department, 16 April 1943, Folder, "OCD Civil Air Patrol," Box 10, Entry 107A, Budgetary Administration Records for Emergency and War Agencies and Defense Activities, 1939–1949 (39.19), RG51, NARA; Mauck, "Civilian Defense in the United States," chap. 9, 11–14; and Earle L. Johnson to John F. Curry, 12 March 1943, Folder 2, Box 2, ELJ, WRHS.

96. J. A. Ulio by order of the Secretary of War, War Department, Adjutant General's Office, memorandum No. W95-12-43, subject: Transfer of Civil Air Patrol from the Office of Civilian Defense to the War Department, 4 May 1943, Binder "Legal Status, Administrative Concepts, and Relationship of the Civil Air Patrol, 1941 to 1949," CAP-NAHC.

97. Civil Air Patrol Coastal Patrol Base No. 16 S-2 Journal, entries for 2 and 10 April, 17 March, and August 16, 1943, Dare County Regional Airport, Manteo, NC; Warner and Grove, *Base Twenty-one*, 83–84; "Panam: Panamanian Motor Tanker," *Uboat.net*, http://uboat.net/allies/merchants/ships/2911.html; and KTB for Seventh War Patrol of *U-129*, entry for 4 May 1943, *U-boat Archive*, http://uboatarchive.net/U-129/KTB129-7.htm.

98. Richard E. Edwards to Adolphus Andrews and James L. Kauffman, memorandum, subject: Civil Air Patrol Coastal Patrol, 18 May 1943, Reel A4064, AFHRA; and Barney M. Giles to B. C. McCaffree, memorandum, subject: Civil Air Patrol Coastal Patrol, 18 May 1943, Folder "SAS 324.3–Civil Air Patrol," Box 93, HHA, LOC.

99. OCD, CAPNHQ, Coastal Patrol Circular no. 49, Expenditures for Permanent Improvements, 15 May 1943, Folder "Civil Air Patrol 1 April 1943–31 May 1943," Box 21, Entry 16A, General Correspondence, 1941–May 1945, Civil Air Patrol, RG171, NARA.

100. War Department, Headquarters Army Air Forces, AAF Regulation no. 20-18, Organization–Civil Air Patrol, 25 May 1943, BLS, CAP-NAHC.

101. Ernest J. King to Adolphus Andrews, memorandum, subject: Civil Air Patrol Coastal Patrol, 30 April 1943, Reel A4064, AFHRA

102. War Department, Headquarters Army Air Forces, AAF Regulation no. 20-18, Organization–Civil Air Patrol, 25 May 1943; Harry H. Blee to H. B. Benedict, memorandum, subject: Special Operational Equipment for Field Activities of Civil Air Patrol, 23 June 1943; C. B. Stone III to Commanding General, Air Service Command, memorandum, subject: Procedure Concerning Civil Air Patrol Supplies, 2 June 1943; Minutes of Conference, 3 June 1943, Patterson Field, Dayton, Ohio, Air Service Command and Civil Air Patrol, BLS; and War Department, Headquarters Army Air Forces, AAF Regulation no. 65-63, Supply and Maintenance–Procedure

for Furnishing Supplies and Services to the Civil Air Patrol, 10 September 1943, Binder "CAP Historical Research–Policy File NR 1 1941-1945," CHF, CAP-NAHC.

103. Earle L. Johnson to All Base Commanders, memorandum, subject: Operation of Airplanes on Coastal and Liaison Patrol Duty, 19 June 1943, BLS, CAP--NAHC.

104. Isaac W. Burnham II to R. E. Schreder, memorandum, subject: Statistics CAP Coastal Patrol No. 4, 28 June 1943, Reel 38920, AFHRA. The two men killed that Burham alluded to were Capt Harry L. Lundquist and FO David S. Williams, 27 June 1943, flying out of Coastal Patrol Base No. 21, Beaufort, NC. Warner and Grove, *Base Twenty-one*, 90–91.

105. See appendix B.

106. George C. Marshall to Ernest J. King, memorandum, 28 June 1943, Folder "A16-3 (9)–Submarine and Anti-Submarine Warfare. File #1 1943," Box 672, Headquarters COMINCH, 1943–Secret, A16-3 (9), RG38, NARA.

107. Craven and Cate, *Europe: Torch to Pointblank*, 389–409; Blair, *The Hunted*, 308–10, 321–22; Schoenfeld, *Stalking the U-Boat*, 166–69; Morison, *Battle of the Atlantic*, 244–47; and US Air Force, *Antisubmarine Command*, 62–84.

108. Earle L. Johnson to CAP Coastal Patrol Personnel, memorandum, subject: Discontinuance of CAP Coastal Patrol, 15 July 1943, BLS, CAP-NAHC.

109. P. C. Growen to Westside T. Larson, memorandum, subject: 3rd Ind., 15 July 1943, Reel A4064, AFHRA.

110. Earle L. Johnson to Coastal Patrol Commanders, memorandum, subject: Suggestions for Liquidation of Certain Coastal Patrol Bases, 16 August 1943, Reel 38920, AFHRA. By January 1944, Bases No. 3, 10, and 11 were still in the process of liquidation. Base No. 6 remained on inactive status pending completion of plans for possible transfer. Of the remaining bases, aircraft and personnel merged to create eight tow target units, four on each coast. These consisted of Tow Target Unit No. 1, New Brunswick, NJ; Tow Target Unit No. 5, Falmouth, MA; Tow Target Unit No. 17, Clinton, MD; Tow Target Unit No. 21, Driver, VA; Tow Target Unit No. 7, Glendale, CA; Tow Target Unit No. 12, San Diego, CA; Tow Target Unit No. 15, San Jose, CA; and Tow Target Unit No. 20, Ft. Lewis, WA. Kendall K. Hoyt to Historical Division, AC/AS, Intelligence, memorandum, subject: Civil Air Patrol, Week Ending 25 December 1943, 24 December 1943; Kendall K. Hoyt to Historical Division, AC/AS, Intelligence, memorandum, subject: Civil Air Patrol, Week Ending 1 January 1944, 31 December 1943; and Kendall K. Hoyt to Historical Division, AC/AS, Intelligence, memorandum, subject: Civil Air Patrol, Week Ended 8 January 1944, 8 January 1944, Folder 4, Box 1, ELJ, WRHS; and CAPNHQ, Operations Orders no. 2, Activation of CAP Tow Target Units, 7 March 1944, Reel 38918, AFHRA.

111. Headquarters, Air Forces, Eastern Defense Command and First Air Force, General Order Number 104, 31 August 1943, Reel A4057, AFHRA. The 25th and 26th Antisubmarine Wings were disbanded on 15 October 1943. Maurer, *Combat Units*, 388–89.

112. CAPNHQ, Summary of CAP Coastal Patrol Operations, 3 September 1943, in CAPNHQ, Report of Civil Air Patrol, 28 December 1943, Folder 4, Box 1, ELJ, WRHS.

113. Table, "Hours Flown by Civil Air Patrol," compiled from Army Air Forces Antisubmarine Command monthly intelligence reports from October 1942 to August 1943, in Antisubmarine Command Historical Section, "CAP History of Operations (First Narrative)," 12 October 1943, Reel A4057, AFHRA.

114. Warner and Grove, *Base Twenty-one*, 100; "Pepper Praises Civil Air Patrol," *Palm Beach Post*, 1 September 1943, 1; "Civil Air Patrol Leaves Its Roanoke Island

Base This Week for Other Posts," *Dare County Times*, 3 September 1943, 1; Boudreau, *CAPCP Base-17*, 51–54; Keefer, *From Maine to Mexico*, 89.

115. "Pepper Praises Civil Air Patrol," *Palm Beach Post*, 1 September 1943, 1.

116. "'Well Done, Good and Faithful,'" *Dare County Times*, 3 September 1943, 2.

117. Adolphus Andrews to Commanding Officer, CAP Coastal Patrol No. 21, memorandum, subject: Services, Appreciation of, 27 August 1943, Folder "#312 – Aeronautics, Jan-Aug 1943," Box 20, Carl Thomas Durham Papers, #3507, Southern Historical Collection, The Wilson Library, University of North Carolina at Chapel Hill, NC.

118. Ernest J. King to George C. Marshall, memorandum, subject: Civil Air Patrol Coastal Patrol, 11 August 1943; Earle L. Johnson to Civil Air Patrol Coastal Patrol Base Commanders, memorandum, subject: Commendation of Civil Air Patrol Coastal Patrol, 31 August 1943, BLS, CAP-NAHC.

119. Remarks of Gen Henry H. Arnold, delivered through transcription at dinners under the sponsorship of the Civil Air Patrol, 16 December 1944, Folder "Speeches and Writings File, 1944," Box 237, HHA, LOC.

Chapter 7

Past Reflections and Future Possibilities

From a very dark eighth of December in 1941 throughout the hectic days until V-J day signified our victory, the personnel of the Civil Air Patrol gave unselfishly of their initiative, time, money and lives, too, to carry out the program which they evolved.

—Gen Henry H. "Hap" Arnold, 7 January 1946

For 18 months, civilian volunteers flew privately owned civilian aircraft over the Atlantic Ocean and Gulf of Mexico in defense of the United States. CAP's coastal patrol effort leveraged the nation's civilian aviation community and transformed private citizens into trained, disciplined, uniformed professionals. Private citizens—with operational guidance from the nation's military—organized, equipped, and operated a coastal patrol effort at 21 independent air bases. With minimal funding from the federal government supplemented by private industry and often the wallets of the volunteers themselves, CAP's coastal patrol service provided a stopgap measure when the nation's armed forces lacked the assets to deter and constrain enemy submarine operations.

As a component of the overall American antisubmarine defense plan, CAP proved senior military leaders wrong in their estimation of civilian volunteers. The disconnect between Army and Navy leadership regarding use of CAP may have slowed the expansion of the coastal patrol effort. Fortunately, key leaders, notably Andrews and Arnold, recognized the value in the light aircraft when the prime doubter, King, felt otherwise. Arguably the pressing need for eyes over the home waters in a deterrent role kept CAP's coastal patrol effort operational despite any interservice disagreement. With funding through the Army, the Navy at least received more patrol aircraft over the shipping lanes at practically negligible investment on its part.

The entire subexperiment easily could have failed in the first weeks. The initial flights could have gotten lost on patrol and required help from the Army or Navy to safely return home. Observers might have struggled to locate and identify items in the water, unable to provide useful intelligence to military personnel. Privately owned aircraft could have crashed out at sea because of mechanical issues or pilot error. These first patrols could have been shot down, intentionally or accidentally. If the volunteer aircrew had been killed or even captured by the enemy, the fallout from the incident may well have immediately ended the subexperiment while handing a propaganda victory to the Germans. None of these fears materialized. The CAP coastal patrol personnel conducted their operations with quiet professionalism. They learned, adapted, and earned the respect of the military establishment.

The success of the subexperiment with the coastal patrol ensured the success of the grand experiment for the entirety of CAP. The small cadre of CAP coastal patrol members proved the essential vanguard to convince Army leadership of CAP's capacity to become a semimilitary auxiliary able to supplement and replace military units needed elsewhere. Civilian volunteers demonstrated sufficient competency and professionalism for the military to equip and entrust CAP members with weapons and release authority to use deadly force against lawful combatants. Furthermore, the civilian effort, with minimal military guidance, managed to sustain and grow operations over the course of its existence. With a motley assortment of low-technology, cloth-skinned aircraft, CAP integrated with increasingly sophisticated military assets. Together, this civil-military aerial umbrella safely escorted thousands of American and Allied merchant ships and their crews along the nation's coastlines, enabling safe passage to ports in the European and Pacific theaters of war.

The untold numbers of men, ships, and war materiel that safely left American factories to defeat the forces of the Axis owed CAP a debt of protective gratitude of dollars, lives, and months of bloody fighting saved. Although armed, the CAP coastal patrols' primary duty was never to destroy submarines. While the organization credited itself in 1943 with damaging or destroying two submarines, postwar records indicate no enemy submarine was damaged, destroyed, or directly attacked by a CAP aircraft.[1]

This brings up the question of the CAP coastal patrol's combat effectiveness. Around 1946, former Base No. 3 pilot Zack Mosley recalled a conversation with Colonel Johnson where the CAP chief recounted interviewing a former U-boat commander. Johnson asked what the commander thought was the most outstanding factor of the defeat off the Atlantic coast. The U-boat man replied: "It was because of those damned little red and yellow planes!"[2] The quotation makes for a great CAP anecdote but is at best apocryphal. One former U-boat captain, Peter-Erich Cremer, led *U-333* to several attacks off West Palm Beach in May 1942 while Mosley's base was actively looking for the enemy. In his postwar memoirs, a bemused Cremer dismissively described CAP: "Patriotic amateur airmen had got together and obtained permission to set up their own flying corps. These unofficial operations increased the confusion by reporting U-boats everywhere, which almost always turned out to be driftwood or portions of wrecks. . . . Though their value was precisely nil the participants had fun, besides receiving a boost to their morale, and had an opportunity to indulge their love of air . . . travel free of charge, with Uncle Sam providing the fuel and food."[3] While Cremer's reflection is partially true, his comment and Johnson's remark mask the increasing influence of technology and airpower in the Battle of the Atlantic.

In the field of technology, the CAP coastal patrol was distinctively limited. Flight Officer David R. Thompson of West Palm Beach flew at Base No. 3 and witnessed the changing nature of antisubmarine warfare at Morrison Field. Beyond the eyes of the pilot and observer, "we were pretty unsophisticated in lacking any sort of sonar or radar or any electronic instrumentation on the planes."[4] The low horsepower engines and small size of the aircraft prohibited adding much weight beyond the ordnance that already strained airframes. In July 1943, the Army investigated the possibility of borrowing a CAP Stinson Voyager 10A to serve as a "guinea pig" for the installation of magnetic anomaly detector equipment but apparently chose not to pursue the matter for undocumented reasons.[5] For the remainder of its operation, the only advanced abilities CAP coastal patrol personnel received came in the form of antisubmarine intelligence for educational purposes.

As originally intended, CAP coastal patrols were flown to inhibit enemy submarines from sinking merchant vessels and to deter attacks off the nation's coasts. In 1927, World War I U-boat veteran Heino von Heimburg, a contemporary of Doenitz, presciently wrote that "air power in the future will increasingly force U-boats to remain submerged by day."[6] By summer 1943, airpower progressively drew first blood, aided tremendously by cryptologic efforts and the development and fielding of aircraft-borne centimetric radar units together with improved underwater listening apparatuses. Through the increased use of long-range bombers and the introduction of escort carriers with convoys, aircraft working alone or in conjunction with surface vessels could now detect and destroy U-boats with increased efficiency. After suffering the devastating loss of 41 operational U-boats in May 1943, including the death of his son, Peter, Doenitz admitted, "we had lost the Battle of the Atlantic."[7] For the remainder of the war, U-boats operated on the defensive, technologically outmatched. Airpower played a decisive role in victory. Forty-five percent of the 648 U-boats lost at sea in the war came through the direct or indirect involvement of aircraft. Over 60 percent of the frontline men who sailed in the U-boat arm died in the war, a figure unequaled by any military service arm or branch in modern war.[8]

Statistically separating out CAP's distinctive contribution to defeating enemy submarine operations from those of the Army, Navy, and Coast Guard is impractical. For example, CAP's 244,600 hours of patrolling the merchant shipping lanes coincided with the implementation of convoys and increased numbers of military aircraft and surface vessels, all patrolling and searching for the enemy. Statistics do not capture the impact of the two most formidable assets in CAP's war against U-boats: the individual coastal patrol aircraft and the eyes of the aircrew. Deterrence is a nebulous matter to objectify into metrics of ships and cargoes saved, but it did cost the lives of 26 civilian volunteers. As part of a larger effort, however, CAP's contribution proved valuable enough to progress from a small experiment to an expanded, sustained operation for a relative pittance of resources. CAP, in conjunction with the nation's armed forces, collectively ensured the safety of the nation's coastal waters in the critical period after entry into World War II.

Amid the accomplishment in sustaining the civilian coastal patrol effort are also considerable shortcomings. CAP National Headquarters failed to address two matters with any expediency: training and logistics. For the former, CAP coastal patrol aircrews did not receive antisubmarine warfare training until May 1943, at which point the U-boat threat for the nation's coasts was essentially over. Logistically, the headquarters never chose to establish any aircraft parts supply depot or move quickly enough to secure the necessary priority rating before the pace of operations and conduct of the war overtook maintenance. This decision can partially be understood in the context of the coastal patrol mission originally being an experiment. Before leaving OCD, Reed Landis addressed the matter of priorities on materiel in a memorandum to Curry. Noting the availability of thousands of civil aircraft not then in use by the armed forces, Landis remarked, "It is our belief that these aircraft should be used so long as they are available. Their maintenance has not yet proven to be any burden on the production of military supplies and it should not prove to be a burden on such production."[9]

By summer 1942, however, the Army began adopting militarized versions of many of the same civilian aircraft favored for coastal patrol duty. Both the Army Air Forces and the Field Artillery pressed light aircraft into observation and liaison roles, notably the UC-61 (Fairchild 24), L-9B (Stinson Voyager) and the purpose-built military variant of the Voyager, the L-5.[10] CAP operated with two mistaken beliefs: First, enough privately owned aircraft could supply operational requirements, and second, CAP's requirements for parts from manufacturers of light aircraft would not interfere with those of the Army. Both beliefs failed to account for the sheer number of flying hours, variety, and scale of CAP's missions.

Replacement aircraft parts proved only part of the equation of operational safety and the lives of the CAP aircrew. CAP's national leadership took a reactive rather than proactive approach to implementation of operational procedures and policies to ensure aircrew safety and rest. To some extent, CAP's personnel were expendable; as civilian belligerents rather than military veterans, the government had no legal obligation to provide any medical treatment, disability, or death benefits. The dearth of replacement parts compounded by a diverse fleet of prewar civilian airframes and engines, together with limited support personnel and maintenance facilities, resulted in dangerous aircraft flown by tired crews. Considering the tedious daily patrols, lack of resources, and ordnance-strained airframes, it is either a miracle or a testament to American aeronautical engineering that only 26 CAP coastal patrol personnel died during the 18 months of operations. Proper lifesaving equipment would have undoubtedly lowered the death toll, but here, too, the equipment issues were a civilian volunteer matter until men began to die and morale declined.

Why the shortcomings at headquarters? A 1944 inspector general survey report by Lt Col Dudley M. Outcalt faulted CAP's senior uniformed leaders. The inspector found CAP National Headquarters disorganized and chaotic, from a military perspective, with low morale. Johnson described his staff as "just a gang of good civilians who have not any of them had army experience."[11] Outcalt's report remarks that when "attempts are made to combine business [civilian] and military practices, as they must inevitably try to combine them, confusion, misunderstanding, and breakdown result."[12] A serious consequence of the entire situation was the failure of headquarters personnel to implement a formal inspection system. Blee, and on occasion Johnson, were the only officials to visit the coastal patrol bases. Blee, however, took no notes and addressed matters case by case. "No technical inspection of the airplanes has been had," wrote Outcalt, and "the only conclusion that can be drawn is that if their accident record is good, as they state, they have been extremely fortunate."[13]

Arguably then, greater initial investment in personnel and dollars in CAP's coastal patrol effort from the Army might have altered U-boat operations off the East and Gulf Coasts. Acknowledging the War Department's investment in CAP, Outcalt concluded that "the Army Air Forces owes a duty to see to it in some way that the equipment is maintained in accordance with reasonable standards."[14] Curry left the headquarters just as the coastal patrol experiment had begun to expand. His replacement as national commander, Johnson, had essentially no military experience. All management of coastal patrol operations fell to Blee. Additional staff or liaison officers for the bases may have provided the strategic foresight to address problems before they threatened the entire operation.

Recognizing the monthly cost to operate a coastal base versus production of a single B-17 or B-24 heavy bomber, a modicum of additional War Department funding could have placed more CAP aircraft over the nation's coasts in areas of high U-boat traffic during the critical period of April–June 1942.[15] Instead, the burden for everything fell on CAP and its small national headquarters to beg, borrow, and cajole its way to standing up patrol bases. This argument particularly applies to Gulf Coast states where hastily erected CAP bases and coastal patrols would have provided air coverage for tankers sailing off the Louisiana coast. Regardless of being armed or not, the deterrent "scarecrow" factor of CAP would have coupled well with Andrew's "bucket brigade" and encouraged German admiral Karl Doenitz and his U-boats to seek easier hunting elsewhere. Dating back to December 1941, the U-boat offensive had sunk 609 ships, fully one-quarter of all Allied shipping Germany would destroy in the entire war.[16]

CAP's coastal patrol experience is a noteworthy success story in the history of American civil-military relations. Volunteers possessed of genuine loyalty and devotion to duty flew their aircraft to the point of critical failure and took considerable risk in service to the military and public.[17] Ironically, at the point when the CAP coastal patrol effort had its most competent personnel along with Army administration and logistical agreements to more adequately maintain its aircraft and bases, the tides of war had turned in the Battle of the Atlantic, negating the need for CAP's services.

The War Department nonetheless recognized CAP as a proven partner in the defense of the home front. In August 1944, the War Department issued a memorandum clarifying CAP's exact status. Under the signature of General Marshall, the memorandum stated that "Civil Air Patrol, an auxiliary of the Army Air Forces, is an official AAF agency."[18] In September, Arnold chose to modify CAP's uniformed appearance to further conform with the Army. He recommended replacing the OCD-required red shoulder loops and sleeve braid. The commanding general opted for this because, now that CAP was under his supervision and performed services on behalf of the country and the Army, the red shoulder loops and sleeve braid did "not properly represent the relationship" between the civilian volunteers and the Army Air Forces.[19]

Belated recognition came to the coastal patrol veterans. Beginning in late 1946, the War Department presented 25 War Department Exceptional Civilian Service Medals to the commanders of the coastal patrol, tow target, and southern liaison patrol bases.[20] In April 1948, the Department of the Air Force awarded 822 Air Medals to CAP coastal patrol aircrew who flew a minimum of 200 hours of over-water patrol time, either as pilots or observers. Sharp and Edwards each received an oak leaf cluster to add to their 1943 Air Medals.[21] The citation accompanying the medal, signed by Pres. Harry S. Truman, listed the recipient and succinctly captured the essence of the award:

> For meritorious achievement while participating in antisubmarine patrol missions during World War II. The accomplishment of these missions in light commercial type aircraft despite the hazards of unfavorable weather conditions reflects the highest credit upon this valiant member of the Civil Air Patrol. The high degree of competence and exceptional courage he displayed in the voluntary performance of a hazardous and difficult task contributed in large measure to the security of coastal shipping and military supply lines. His patriotic efforts aided materially in the accomplishment of a vital mission of the Army Air Forces in the prosecution of the war.[22]

For all other CAP personnel who participated in the coastal patrol, tow target, or southern liaison patrols, in April 1949, the Air Force issued Certificates of Honorable Service recognizing the named recipient as having served on CAP active duty as a belligerent with the Armed Forces of the United States during World War II."[23]

Analysis of CAP's coastal patrol experience leads to a series of conclusions critical to discussing the future employment of the organization in domestic emergency or wartime contingencies. These conclusions are oriented to maximize the potential impact of CAP resources for airpower practitioners and are framed broadly to provide flexibility for future potentialities. Perhaps then these conclusions are best considered to be doctrinal in nature.

> Aided by access to military training resources, auxiliary Airmen can function as semi-military professionals.
>
> Auxiliary Airmen can supplement or replace uniformed military personnel in certain domestic roles.
>
> Auxiliary Airmen can provide a similar or near-similar product to military commanders at a reduced overall cost.
>
> Auxiliary Airmen, volunteering their time, resources, and lives, will faithfully serve alongside uniformed personnel when their services are requested.
>
> Auxiliary Airmen can operate in a joint command environment.
>
> Strong civil-military relations with clear lines of communication are critical for maximizing the potential impact of the auxiliary Airmen.
>
> Auxiliary aviation resources are capable of sustained, high tempo operations only with the infusion of financial and logistical resources.
>
> Auxiliary Airmen can incorporate and operate sophisticated military technologies with civilian aircraft.
>
> Without financial or logistical resources, auxiliary Airmen are best employed for specific, temporary tasks.

The successful civil-military relationship embodied in the coastal patrol effort is the critical factor in CAP's postwar existence. The contemporary CAP organization, however, is grounded in legislation passed into law in the immediate aftermath of World War II, which has charted a peacetime course of action. As early as November 1944, Johnson wrote to Arnold about the postwar plan for CAP. He credited CAP's success as "due largely to its status as an auxiliary of the AAF," with CAP's reputation "gained through the Coastal Patrol and other active missions for the AAF." Further acknowledging how "the day of active military missions is nearly done," Johnson's postwar plan for CAP focused on continuing most of the organization's present activities to continue as the Army Air Forces' auxiliary.[24] From 10–11 January 1946, Generals Arnold and Carl A. Spaatz informed

CAP leadership of the forthcoming termination of the present national emergency and the legal authority for the continuation and financial support of CAP. Although funding would cease on 31 March, both Army Air Forces leaders pledged assistance in obtaining a federal charter for CAP.[25]

In the aftermath of the two generals' speeches, CAP's wing commanders organized themselves to draft a future plan. These 48 men voted to appoint a Committee on Post-War Organization under the chairmanship of Arkansas Wing Commander, Lt Col Rex P. Hayes, to study and prepare a plan and recommendations for the fate of the CAP.[26] After meeting from 11–13 February 1946 at Army Air Forces headquarters in Washington, the committee plan for a permanent CAP civilian organization accepted a peacetime goal "to advance aviation to help prevent another war."[27] As the Army Air Forces Auxiliary, CAP would place priority on meeting the needs of the Army Air Forces while promoting civil aviation and continuing as a flying organization in the performance of active missions including search and rescue, disaster relief, mercy missions, forest and flood patrols, and cooperation with law enforcement agencies.

To complement the flying missions, the committee recommended efforts to develop auxiliary communications networks, a rifle and pistol marksmanship program, first aid medical training, physical fitness, and civil aviation improvements.[28] From 27 to 28 February and 1 March, the 48 wing commanders reconvened to consider the committee report, which was approved in its entirety. The wing commanders in turn shared the committee plan with Congressman Sumners, chair of the House Committee on the Judiciary.[29]

On 12 March, Sumners introduced legislation for incorporation, and the Senate Judiciary Committee held hearings on the legislation on 14 May. Maj Gen Frederick Anderson, Assistant Chief of Staff, Army Air Forces, gave a statement to the committee regarding the legislation. Anderson shared his belief that, in the event of future national emergencies, "the better trained, the better equipped, and the better organized this group of civilian aviation personnel may be, the better this country will be able to meet the attacks of an aggressor."[30] His statement alluded to a potential future where CAP would again be called upon to lend itself to the defense of the nation. Yet when President Truman signed Public Law 79-476 into law on 1 July, CAP the corporation listed its objects and purposes as "being solely of a benevolent character."[31] Armed CAP operations passed into history.

As noncombatants, CAP nonetheless could offer a tremendous service to all in need. To further enhance CAP's position in relation to the nation's military, a second key piece of legislation was required. On 26 May 1948, Truman signed Public Law 80-557, establishing CAP as the auxiliary of the newly independent United States Air Force.[32] This legislation authorized the Secretary of the Air Force to extend aid to CAP in the fulfillment of its objectives in the form of aircraft, aircraft parts, and other Air Force equipment. Arguably of greater importance, the legislation also authorized the secretary "in the fulfillment of the noncombatant mission of the Air Force Establishment to accept and utilize the services of the Civil Air Patrol."[33] Together with Public Law 79-476, the collective legislation cemented CAP as a volunteer auxiliary of the Air Force, available for noncombat programs and missions with taxpayer funding and resources available to supplement those resources provided by the members and requisite states.

Reflecting upon CAP's coastal patrol experience and the eight broad conclusions, the future use of CAP by airpower practitioners is unquestionably greater today than in early 1942. As detailed in Air Force Instruction 10-2701, *Organization and Function of the Civil Air Patrol*, Air Force doctrine recognizes CAP members as Airmen when performing missions or programs in the Air Force Auxiliary capacity and as part of the total force in conjunction with the Air Force, Air National Guard, and Air Force Reserve. Under operational direction of First Air Force, CAP functions as the Air Force Auxiliary when its services are used by any department or agency in any branch of the federal government. The secretary of the Air Force, or a designee, authorizes and assigns CAP missions and programs. Every operational Air Force–assigned mission to CAP is executed under military command and control. Although these missions do not involve combat or combat operations, CAP currently provides limited combat training support. The Air Force manages support for CAP in its auxiliary status via funding, equipment, coordination, and integration management.[34]

Compared to the haphazard coastal patrol origins, the contemporary CAP is equipped, trained, and engaged in several sophisticated missions. CAP aircraft regularly participate in Falcon Virgo and Fertile Keynote air defense intercept missions, among others, across the country.[35] Other CAP personnel participate in Operation Green Flag where specially modified Cessna aircraft incorporate L3Harris Wescam

MX-15 multispectral imaging systems to function as surrogate remotely piloted aircraft (RPA). These aircraft support RPA training programs of the Department of Defense and coalition personnel at significantly reduced cost.[36] CAP aircraft work closely with the Air National Guard escorting MQ-9 Reapers for training operations from two to six days a week.[37] CAP's National Radar Analysis and Cell Phone Forensic Teams use proprietary software to convert raw digital data into actionable search information, which has saved hundreds of lives across the country.[38] To increase CAP's imaging capabilities for FEMA and local emergency management agencies, several aircraft are increasing use of the WaldoAir XCAM Ultra50 sensor pod system to provide multispectral imaging capable of producing 3-D models of storm-damaged areas.[39]

Figure 29. A CAP Cessna 182T of the Congressional Squadron is intercepted during a training exercise by an F-16 of the 113th Wing of the DC Air National Guard, 3 October 2016. (Photograph by John Swain via CAP National Headquarters.)

Commencing in 2018, CAP established a full-time program to train and field small unmanned aerial systems (sUAS) for use in post-disaster response missions, rescue efforts, damage assessment, and general aerial reconnaissance. By the end of fiscal year 2019, CAP fielded over 1,500 sUAS units in operation registered with the Federal

Aviation Administration, the largest operator of these systems in the country. The first use of sUAS drones to aid in search and rescue missions commenced in September 2019.⁴⁰ CAP is already engaged at two Air Force bases in counter-sUAS missions and is expanding to other bases. Presumably the discussions between CAP and the Air Force will examine what legal requirements are necessary to maximize the potential of this new and potent tool in the hands of auxiliary Airmen.[41]

This wealth of resources is currently employed in the homeland for noncombatant roles as determined by CAP's federal charter. As noted by Air Force Lt Col Jeremy Hodges, CAP's role as the Air Force's auxiliary "makes a significant contribution to national security strategic objectives," and he advocates further exploration of the use of auxiliary forces to secure national objectives.[42] But what about CAP's use in time of declared war or national emergency as referenced by Anderson? The need for CAP to carry arms and or use deadly force is highly unlikely. There are potential missions where CAP may play a more active role in national defense.

Framed through the case study of coastal patrol, CAP personnel and assets are ideally suited to supplement and replace military personnel. In the immediate, CAP should continue to focus on missions within the United States and its territories by providing support to First Air Force (Air Forces Northern, 1AF [AFNORTH]). Several potential roles where CAP could operate under the current federal charter as tasked by AFNORTH include:

- flying unmanned (sUAS) missions that utilize assorted imaging and information sensors on patrol duty along the nation's borders to observe and report unusual activity;
- using sUAS to fly cellular payloads over remote or adversely affected locations to provide functionality for personal communications devices of ground personnel,
- providing military and domestic agencies with auxiliary radio communication capability through the existing ground-based and aerial repeater network independent of cellular networks;
- conducting light courier and supply missions delivering personnel and equipment of a benevolent nature, such as medical supplies, noncombatant personnel, or humanitarian aid;

- providing supplementary cybersecurity assistance to domestic state and federal agencies; and
- conducting photographic reconnaissance and patrol missions over key civilian facilities and infrastructure.

Under the existing legislation, CAP is investigating new technologies and upgrading existing systems. John W. Desmarais Jr., CAP's national director of operations, is currently exploring acquisition and fielding of higher end sensor packages including light detection and ranging (LIDAR) and multispectral capabilities. For existing search and rescue operations, CAP is pursuing infrared and thermal sensor capabilities, the latter of marked importance as global climate change has increased the risk of wildfires in the Western United States. "Imagine being able to pop a drone over a remote area of dense forest at night," explains Desmarais, "and being able to find a missing person on the ground and then using a sUAS equipped with lights and speakers . . . to tell a person on the ground to stay where they are" with help on the way. Work is also underway to equip additional aircraft with the MX-15 or similar sensors to match Air Force capabilities with a vision of providing realistic training for joint terminal air controllers and Air Force air support operations squadrons. Distributed per region around the country, these CAP aircraft could support both warfighter training and defense support of civil authorities high availability disaster recovery missions.[43]

CAP's growing involvement in the fields of cybersecurity and cyber defense offers the Air Force a vast pool of talent and experience to leverage in times of peace and crisis. CAP's cyber mission began to develop future generations of cyberspace defense operators. Since 2009, CAP cadets have participated in the Air Force Association's Cyber-Patriot national youth cyber education program, steering young men and women to military service in cybersecurity as well as in science, technology, engineering, or mathematics. With a recent partnership between CAP and Cisco Networking Academy, the partnership will provide all volunteer members—cadet and adult members alike—free access to cybersecurity support, training, and curriculum resources. As CAP's cyber defense mission continues to grow, should or when a potential mass cyberattack strike the homeland, CAP members can augment existing Air Force or state and local government cybersecurity efforts to maintain or defend critical services and digital infrastructure.[44]

Figure 30. In 2018, CAP Cessna 182T aircraft employed specialized WaldoAir camera systems to capture multispectral imagery of damage in North Carolina caused by Hurricane Florence. (Photograph courtesy of CAP National Headquarters.)

However, should CAP's charter be amended before or after a declaration of war to authorize the Secretary of the Air Force to employ auxiliary forces for nonbenevolent missions, the potential missions of CAP are great. CAP volunteers could engage in the following roles:

- Flying tactical airlift missions moving combatant personnel and or select military supplies and equipment
- Supplementing or replacing uniformed Air Force personnel in select domestic air base duties as professional education/certification and necessity requires

- Assisting as trainers for aviators and military personnel
- Serving as force multipliers in cyber defense and information assurance for federal, state, and local partners
- Flying manned or unmanned reconnaissance missions to gather real-time information for intelligence, surveillance, and reconnaissance purposes

For a future conflict with an unknown adversary, the Air Force can potentially draw upon CAP to execute these and other missions. Considering the improbability of a conventional enemy land force invading the continental United States, physical CAP assets will assist the Air Force along the nation's borders, in cyberspace, and throughout the interior. The need to arm CAP aircraft and aircrew with conventional weapons is unlikely. Future defensive weaponry will probably entail a use of cyber assets and intelligence gathering in concert with armed uniformed military forces and civilian law enforcement.

Possible scenarios for peacetime training or military operations include leveraging CAP aerial assets in numbers, whether sUAS to produce impromptu drone swarms or mobilizing the manned aircraft for resupply, deterrent, and reconnaissance missions. Using light aircraft in the role of forward air controllers (FAC) has served the Army and Air Force nobly throughout the twentieth century. Likewise, CAP's surrogate RPAs could be expanded in number with relatively minor modification to the airframes to carry either military or CAP personnel in a FAC role, coordinating with American or coalition forces on the ground and armed Air Force and Navy strike fighters aloft.

In terms of CAP personnel, several actions can be taken today to better position the auxiliary as a force multiplier for contemporary and future Air Force needs. The most immediate action is to leverage modern data analytics to track the professional certifications and skill sets of auxiliary Airmen.[45] The contemporary costs of gathering of CAP member data are negligible compared to the potential benefits of the data. For example, while CAP circa 1942 sought experienced pilots and A&E certified mechanics for the immediate needs of the coastal patrol, CAP circa 2020 would also want to know of members with professional certifications in the fields of cybersecurity, medicine, law, engineering, information technology, and more.

Another obvious skill set are those CAP members who are retired or separated military personnel. These individuals, barring disqualifying characteristics, would be ideal to fill domestic active military positions, thereby releasing active, guard, and reserve Airmen for deployment. While the capability now exists to convert CAP aircraft into unmanned drones, emphasis must be placed on the human operator value added by auxiliary Airmen if and when peer and near-peer adversaries disrupt or disable drone comunications.[46]

Figure 31. In Syracuse, New York, CAP Cessnas have provided escort for New York Air National Guard MQ-9 Reaper remotely piloted aircraft flying to and from restricted air space. (Photograph by Leslie Vazquez via CAP National Headquarters.)

A second action is to better integrate auxiliary Airmen—where certification and qualifications permit—with their total force partners. Currently there are ongoing interactions of either training or emergency response nature. As with the coastal patrol effort, the personnel involved are a mere fraction of the auxiliary membership. Leveraging the existing wing and region structure, CAP should pursue discussions with Air National Guard and Air Force Reserve components within the respective states to see where Airmen could more closely work and interact together to form stronger intra- and interprofessional relationships between the elements of the total force and between the Air Force and the civic community. The objective of both suggested actions is to improve communication with, awareness of, and trust in CAP's volunteers and their skills, which are available to local, state, and federal agencies in the event of extraordinary circumstances.

Third, CAP, acting as a bridge for civil-military relations, is uniquely positioned to serve as a coordinating agency for the employment of light general aviation in contingency operations. In a future declared war or national emergency, CAP can and should serve as the federal coordinating element for general aviation operations. With the right regulatory framework between the Federal Aviation Administration and the Air Force, CAP could oversee an additional volunteer force of light civilian aircraft specifically employed for emergency response. These small civilian aircraft could function as light tactical airlift for evacuation, observation, and possibly even medevac roles, something akin to the aircraft of the Civil Reserve Air Fleet.[47] For FEMA and state emergency management agencies, CAP could theoretically coordinate a fleet of light aircraft, including pilots with appropriate capabilities, in mere hours rather than shuttle CAP corporate aircraft from neighboring states over the course of several days.

A fourth action, although one requiring substantial federal investment, is to take preventative measures to better prepare physical CAP assets for potential military use. A potential electromagnetic pulse (EMP) attack from a high-altitude nuclear explosion would severely cripple electronic systems in affected areas. By hardening physical equipment in CAP's extensive communications network against EMP, the Air Force would equip itself and state emergency management agencies with a potential auxiliary asset, freeing vital military systems for countering additional attacks on the homeland. Hardening of sensitive electronics in CAP's fleet of aircraft and ground vehicles would likewise ensure physical auxiliary assets are available to supplement or replace active, Guard, and Reserve assets engaged in civilian emergency operations for essential offensive or defensive military operations.[48]

Much as the situation in the Battle of the Atlantic found the fledging CAP pressed into antisubmarine operations, any future military conflict will by necessity be of sufficiently grave threat to the homeland to require the auxiliary to undertake nonbenevolent operations. The eight conclusions drawn from the CAP coastal patrol experience apply to any and all future employment of CAP as the auxiliary of the Air Force. A ninth conclusion can also be drawn from CAP's 18-month long armed experience: mainly how volunteers will, by necessity, find creative and practical solutions to operational problems. As a member of the total force, CAP's resource base is strong and the membership faithful and eager to serve as the auxiliary of the Air Force.

Figure 32. A Cessna 172P of the Connecticut Wing participates in Operation Bird Dog on 5 May 2020. CAP aircraft carried Navy observers in the joint exercise with Naval Base New London, Groton, CT. The exercise demonstrated CAP capabilities in aerial antiterrorism force protection support while also training new submarine commanders through an opposing force simulated event. (Photograph by George D. Stewart via CAP National Headquarters.)

At the time of this book's publication, CAP has been actively engaged in missions in response to the COVID-19 pandemic. In the corporation's largest sustained national effort since World War II, CAP has provided a "crucial mainstay" to First Air Force operations in responding to the pandemic. CAP volunteers have staffed emergency operation centers, delivered thousands of pounds of personal protective equipment, delivered test kits to hospitals and samples to labs by air and ground, and supported state and federal agencies in an array of tasks.[49] If CAP is called upon in time of declared war or national emergency, the challenges it confronts will be creatively addressed by the civilian volunteers in partnership with the Air Force to meet the disaster response and security needs of the homeland.

Notes

Epigraph. Henry H. Arnold to Frank E. Dawson, memorandum, subject: Letter of Appreciation, 7 January 1946, courtesy of the Charles Small Family, Richmond, VA.

---

1. See appendix F.
2. Mosley, *Brave Coward Zack*, 68; and Neprud, *Flying Minute Men*, 46.
3. Cremer, *U-Boat Commander*, 78.
4. Thompson and Thompson, *Palm Beach*, 151.
5. Harold M. McClelland to Edwin E. Aldrin, memorandum, subject: Airplane for Anti-Submarine Activity, 23 July 1942; Harold M. McClelland to Earle L. Johnson, memorandum, subject: 1st Indorsement, 23 July 1942; Earle L. Johnson to Edwin E. Aldrin, memorandum, subject: Civil Air Patrol planes on Coastal Patrol Duty, 21 July 1942; Harold McClelland to Edward L. Bowles, 4 August 1942, Folder "400.01 Dr. Bowles's SADU Papers," Box 6, Entry 117, Secretary of War, Office, Expert Consultant to the Secretary of War, Correspondence and Reports, Re: Edward L. Bowles, Anti-Sub Warfare, RG107, NARA.
6. Mulligan, *Neither Sharks nor Wolves*, 58.
7. Doenitz, *Memoirs*, 341.
8. Roskill, *The Period of Balance*, vol. 2 of *The War at Sea*, 351–79; Blair, *The Hunted*, 311–53, 710–11; Overy, *Why the Allies Won*, 45–62; Gannon, *Black May*; Mulligan, *Neither Sharks nor Wolves*, 70–88, 256; and Niestlé, *German U-boat Losses During World War II*, 3–4, 198–99.
9. Reed G. Landis to John F. Curry, memorandum, 13 March 1942, BLS, CAP-NAHC.
10. Holley, *Army Air Forces Historical Studies no. 44*, 78–109; Raines, *Eyes of the Artillery*, 31–83, 327–31; Swanborough and Bowers, *United States Military Aircraft Since 1909*, 574–75, 644; Ten Eyck, *Jeeps in the Sky*, 15–23, 37–60; and Love, *L-Birds*, 35–36, 44, 47–48.
11. War Department, Headquarters of the Army Air Forces, Dudley M. Outcalt to Air Inspector, memorandum, subject: Survey of the Civil Air Patrol, 8 March 1944, 18, Folder 3, Box 5, ELJ, WRHS.
12. War Department, Headquarters of the Army Air Forces, Dudley M. Outcalt to Air Inspector, memorandum, subject: Survey of the Civil Air Patrol, 8 March 1944, 42, Folder 3, Box 5, ELJ, WRHS.
13. War Department, Headquarters of the Army Air Forces, Dudley M. Outcalt to Air Inspector, memorandum, subject: Survey of the Civil Air Patrol, 8 March 1944, 42, Folder 3, Box 5, ELJ, WRHS.
14. War Department, Headquarters of the Army Air Forces, Dudley M. Outcalt to Air Inspector, memorandum, subject: Survey of the Civil Air Patrol, 8 March 1944, 42, Folder 3, Box 5, ELJ, WRHS.
15. A Boeing B-17 Flying Fortress cost from $301,000 (1939–41) to $188,000 (1945). A Consolidated B-24 Liberator cost from $379,000 (1939–41) to $215,500 (1944). As of July 1943, the monthly cost for a CAP coastal patrol base averaged $29,657. Hyde, *Arsenal of Democracy*, 114; Louis F. Licht, Jr. to Henry H. Arnold, memorandum, subject: Cost of CAP Coastal Patrol, July 1943, 8 October 1943, and inclosure, "Expenses of In-Shore Anti Submarine Patrol Conducted by Civil Air Patrol during Month of July, 1943," 6 October 1943, Reel 38920, AFHRA.
16. Blair, *U-Boat War*, 590–95, 691–700; Craven and Cate, *Plans and Early Operations*, 533–35; Roskill, *The Period of Balance*, vol. 2, *The War at Sea*, 102–7; and Doenitz, *Memoirs*, 250–52.

17. Earle L. Johnson to Henry L. Stimson, 28 July 1942; and Earle L. Johnson to All Coastal Patrol and Liaison Patrol Commanders, memorandum, subject: Loyalty and Devotion to Duty of Coastal Patrol and Liaison Patrol Personnel, 21 May 1943, BLS, CAP-NAHC.

18. War Department, memorandum no. 95-44, subject: Civil Air Patrol – Army Air Forces Auxiliary, 23 August 1944, Binder "Legal Status, Administrative Concepts, and Relationship of the Civil Air Patrol, 1941 to 1949," CAP-NAHC. The earliest statement indicating CAP had any auxiliary status can be found in CAPNHQ Operations Directive no. 7 of 12 March 1942, listing CAP "as a Volunteer Corps serving as an auxiliary to the Armed Forces." CAP's transfer from OCD to the War Department led to further confusion as to its auxiliary status prior to the issuance of memorandum no. 95-44. OCD, CAPNHQ, Operations Directive no. 7, Rules of Land Warfare, 12 March 1942, Reel 38907, AFHRA; Jack Vilas to Henry A. Hawgood, 24 March 1944; and Henry A. Hawgood to Jack Vilas, memorandum, subject: 1st Indorsement, 28 March 1944, BLS, CAP-NAHC.

19. Henry H. Arnold to Deputy Chief of Staff, memorandum, subject: Re--designation of Civil Air Patrol Uniform, 19 September 1944, Box 167, HHA, LOC. CAP National Headquarters announced removal of the red shoulder loops and sleeve braid on 27 October 1944. A new shoulder sleeve insignia came into use in November which added a red rocker above the previous insignia with the words "CIVIL AIR PATROL" centered in white. CAPNHQ, *CAP Bulletin* 3 no. 32, 27 October 1944, and no. 35, 10 November 1944, Folder 6, Box 6, ELJ, WRHS.

20. Colby, *This Is Your Civil Air Patrol*, 5; "CAP Officer Cited," *Monitor* (McAllen, TX), 19 November 1946, 1; "General Recommends Germany Be Made 'Fool's Work' Example," *Nebraska State Journal* (Lincoln), 25 September 1947, 10; "Highest Civilian Award to Civil Air Patrol Units," *Amarillo Globe-Times* (TX), 23 September 1947, 9; "4 CAP Members Get Army Awards," *Philadelphia Inquirer*, 9 April 1947, 3; "CAP Commanders Get Army Medals," *Palm Beach Post* (West Palm Beach, FL), 20 November 1946, 2; and "Medal Awards Slated Today," *Charlotte Observer*, 26 January 1947, 6.

21. Hopper, *Civil Air Patrol Historical Monograph Number 2*, chap. 3, 9–10; and 32nd Air Force Base Unit, National Headquarters, Civil Air Patrol, Proceedings of the Civil Air Patrol Board 16–17 February 1948, 20 February 1948, 2, CAP-NAHC. There was a discussion to award the American Theater Ribbon to the coastal patrol and southern liaison patrol aircrews. As they were civilians, the Adjutant General clarified they were only eligible to receive the Civilian American Theater Ribbon. Said ribbon was never awarded to any CAP personnel. E. O'Donnell Jr. to F. S. Richards, memorandum, subject: Award of American Theater Ribbon to Certain Civil Air Patrol Members, 8 October 1946; and F. S. Richards to Carl A. Spaatz, memorandum, subject: 1st Indorsement, 31 October 1946, Binder "CAP Historical Research, Policy File NR 2 1946-1947," CHF, CAP-NAHC.

22. Citation to Accompany the Award of the Air Medal to First Lieutenant Henry E. Phipps, 3-3-21, Civil Air Patrol, Henry E. Phipps Papers, CAP-NAHC.

23. Headquarters and Headquarters Squadron, Civil Air Patrol–United States Air Force, Bolling Air Force Base, Lucas V. Beau to CAP Member, 11 April 1949; and Department of the Air Force, Certificate of Honorable Service to First Lieutenant Robert Lee Moore, 4-4-982, Folder "Awards and Certificates of Service," Box 139, Robert L. Moore Papers, Private Collections, World War II, Military Collection, North Carolina State Archives, Raleigh, NC. Despite being issued in April 1949, all the Certificates of Honorable Service are dated 15 May 1948.

24. Earle L. Johnson to Henry H. Arnold via Assistant Chief of Air Staff, Training, memorandum, subject: Postwar Program for Civil Air Patrol, 21 November 1944, Folder 4, Box 1, ELJ, WRHS.

25. Message from General Arnold Read at Civil Air Patrol Conference by Col [Robert] Proctor, 10 January 1946; and Speech of General Spaatz for Civil Air Patrol Conference, 10 January 1946, Binder "Legal Status, Administrative Concepts, and Relationship of the Civil Air Patrol, 1941 to 1949," CAP-NAHC.

26. The committee, chaired by Hayes, consisted of 14 CAP lieutenant colonels, all wing commanders: D. Harold Byrd, TX; Gordon A. DaCosta, IL; Everett Davis, ID; Frank E. Dawson, NC; Edward R. Fenimore, MD; Guy P. Gannett, ME; Lewis W. Graham, NM; Don J. Johnston, IA; Roy W. Milligan, MT; J. Michael Morris, AZ; Bertrand Rhine, CA; George A. Stone, OH; William C. Whelan, TN; Walker W. Winslow, IN. Harry H. Blee to All CAP Unit Commanders, memorandum, subject: Report of Wing Commanders' Conference, 15 January 1946, Binder "Legal Status, Administrative Concepts, and Relationship of the Civil Air Patrol, 1941 to 1949," CAP-NAHC.

27. Harry H. Blee to All CAP Wing Commanders, memorandum, subject: Post-War Plan for Civil Air Patrol, 15 February 1946, with attachment, "Proceedings of the CAP Committee on Post-War Organization and Program," Binder "Legal Status, Administrative Concepts, and Relationship of the Civil Air Patrol, 1941 to 1949," CAP-NAHC.

28. Harry H. Blee to All CAP Wing Commanders, memorandum, subject: Post-War Plan for Civil Air Patrol, 15 February 1946, with attachment, "Proceedings of the CAP Committee on Post-War Organization and Program," Binder "Legal Status, Administrative Concepts, and Relationship of the Civil Air Patrol, 1941 to 1949," CAP-NAHC.

29. Harry H. Blee to All CAP Unit Commanders, memorandum, subject: PostWar Plan for Civil Air Patrol, 4 March 1946, with attachment, "Proceedings of Conference of Wing Commanders of Civil Air Patrol," 4 March 1946, Binder "Legal Status, Administrative Concepts, and Relationship of the Civil Air Patrol, 1941 to 1949," CAP-NAHC.

30. Senate Committee on the Judiciary, *To Incorporate the Civil Air Patrol: Hearings on H.R. 5744*, 79th Cong., 2nd sess., 1946, 5.

31. *Act to Incorporate the Civil Air Patrol*, Public Law 79-476, *US Statutes at Large* 60 (1946): 346–47, codified at *US Code*, Title 36, chap. 403 (2000).

32. Harry S. Truman to George A. Stone, telegram, 26 May 1948, Folder, "Civil Air Patrol," Box 1818, OF 1285-D, Air Force, U.S.–Misc. (1947–Mar. 1950), Official File, Papers of Harry S. Truman, Harry S. Truman Presidential Library and Museum, Independence, MO.

33. *An Act to Establish the Civil Air Patrol as a Civilian Auxiliary of the United States Air Force*, Public Law 80-557, *US Statutes at Large* 62 (1948): 274–75, codified at *US Code*, Title 10, chap. 909 (2000).

34. Air Force Instruction 10-2701, *Organization and Function of the Civil Air Patrol*, 7 August 2018, 3–7, 11.

35. Fertile Keynote missions are practice intercept missions with the CAP aircraft as the target; Falcon Virgo missions are designed to test airspace security with CAP aircraft conducting stimulated aerial attacks. Mary Beth Sheridan, "Civilian Pilots Provide Target Practice," *Washington Post*, 17 January 2007; and Andrea F. Rhode, "Intercept Missions Increase Pilot Readiness," 115th Fighter Wing Public Affairs, Air National Guard, 11 September 2016, https://www.ang.af.mil/Media/Features/Article/1015893/intercept-missions-increase-pilot-readiness/. Other air defense intercept training and evaluation missions involving CAP aircraft include Felix Key-

note, Felix Hawk, Falcon Hawk, Fertile Hawk, and participation in named exercises such as Gunfighter Flag, KOA Shield, and Sentry Shield.

36. Dan Gettinger, "The Surrogate Predator Program," Center for the Study of the Drone, Bard College, 25 January 2016, https://dronecenter.bard.edu/the-surrogate-predator-program/; and Stephen Joiner, "Cessnas Pretending to be Predators," *Air and Space Magazine*, June 2016, https://www.airspacemag.com/flight-today/soundings-cessna-180959131/. CAP's Green Flag program supports combat training from base operating locations in Alexandria, LA (CAP Green Flag East), and Las Vegas, NV (CAP Green Flag West). From CAP Green Flag East, approximately 35,000 US and coalition forces receive training benefit annually. As of August 2019, a total of more than 220,000 US and coalition forces have received training benefit from CAP's Green Flag East support. From CAP Green Flag West, approximately 45,000 US and coalition forces receive training benefit annually. As of August 2019, a total of more than 365,000 US and coalition forces have received combat training benefit from CAP's Green Flag West support. Louis Piccotti, Headquarters, CAP, Senior National Patrol Manager, Special Missions, Maxwell AFB, AL, to the author, email, 28 August 2019.

37. CAP aircraft currently escort Reapers out of Ellington Field Joint Reserve Base, Houston, Texas, and Hancock International Airport, Syracuse, New York. Rick Moriarty, "Mystery of Planes Circling Over East Side of Syracuse Solved," *Syracuse Post-Standard*, 2 September 2017, https://www.syracuse.com/news/2017/09/mystery_of_planes_circling_over_east_side_of_syracuse_solved.html; CAP National Headquarters, "CAP Making Annual Trek to Capitol Hill for Legislative Day," CAP News, 26 February 2019, https://www.cap.news/cap-making-annual-trek-to-capitol-hill-for-legislative-day/; and Rick Moriarty, "Reaper Drones Make History, Fly Unescorted in and out of Syracuse Airport," *Syracuse Post-Standard*, 25 September 2019, https://www.syracuse.com/news/2019/09/reaper-drones-make-history-fly-unescorted-in-and-out-of-syracuse-airport.html.

38. CAP, *2018 Report to Congress*, 19–22, 28; and Vicky Travis, "Cell Phone Forensics on Search Frontlines," *Civil Air Patrol Volunteer* (Fall 2017): 23–26.

39. "Specially Equipped CAP Planes Conducting Florence Damage Assessment Flights," *CAP News*, 20 September 2018, https://www.cap.news/specially-equipped-cap-planes-conducting-florence-damage-assessment-flights/; "CAP Reconnaissance Flights Providing 3-D Views of Michael's Damage," *CAP News*, 9 November 2018, https://www.cap.news/cap-reconnaissance-flights-providing-3-d-views-of-michaels-damage/; and Vicky Travis, "CAP-tested WaldoAir Camera System Provides Interactive 3-D Images," *Civil Air Patrol Volunteer* (Spring 2019): 20–23.

40. CAP, *2019 Annual Report*, 36; Russell Slater, "Small Unmanned Aerial: A Future Trend in Aerial Observation," *Civil Air Patrol Volunteer* (Spring 2018): 16–18; Jennifer S. Kornegay, "CAP's Small UAS Program: It's the Future," *Civil Air Patrol Volunteer* (Spring 2019): 35-37; Desmarais to the author, email, 25 August 2019; and Reese Rezac, "CAP Debuts sUAS as Search and Rescue Tool in S. Dakota," *CAP News*, 10 October 2019,https://www.cap.news/cap-debuts-suas-as-search-and-rescue-tool-in-s-dakota/?fbclid=IwAR0_ylzsw5fRmQoogO7yZM2m8xmjexMEN9Mmsd9rpE8xZJg9bFvRMtI-0cc.

41. Mark Smith, National Commander's Briefing, 2019 CAP National Conference, 9 August 2019, Baltimore, MD.

42. Jeremy K. Hodges, "Using Auxiliary Forces to Accomplish Strategic Objectives," *Air & Space Power Journal* 32, no. 2 (Summer 2018): 55.

43. Desmarais to the author, e-mail.

44. "CAP Cyber Missions," Civil Air Patrol Cadet Programs, https://www.gocivilairpatrol.com/programs/cadets/activities/cyber; "What Is CyberPatriot?," Air

Force Association's CyperPatriot, https://www.uscyberpatriot.org/home; and Jacob Bixler, "CAP Announces Partnership with Cisco Networking Academy," *CAP News, 11 September 2020,* https://www.cap.news/cap-announces-partnership-with-cisco-networking-academy/.

45. The wartime CAP Application for Enlistment included a section asking applicants to "specify other skills you have acquired through study, hobbies, construction of instruments, etc. (for example, photograph, public speaking, cryptanalysis, etc.)." OCD, CAP, "Application for Enlistment Civil Air Patrol," Folder "507 Civil Air Patrol," Box 111, Entry 10, General Correspondence, 1940–1942, 502 to 511, RG171, NARA. CAP chaplains are currently the only volunteers who may be utilized by Air Force personnel, specifically wing chaplains to provide support during contingencies or local emergencies. This is restricted to only exceptional cases where Regular Air Force or reserve component chaplains are unavailable. Air Force Instruction 52-101, *Chaplain, Planning and Organizing,* 15 July 2019, 25–26.

46. Rachel S. Cohen, "Cessna-Turned-Drone Foreshadows Future Unmanned Opportunities," *Air Force Magazine,* 16 August 2019, http://www.airforcemag.com/Features/Pages/2019/August%202019/Cessna-Turned-Drone-Foreshadows-Future-Unmanned-Opportunities.aspx.

47. US Department of Transportation, "Civil Reserve Air Fleet Allocations," 6 May 2019, https://www.transportation.gov/mission/administrations/intelligence-security-emergency-response/civil-reserve-airfleet-allocations.

48. In the 2004 executive report of the Commission to Assess the Threat to the United States from Electromagnetic Pulse (EMP) Attack, the commission examined the feasibility and cost of hardening select military and civilian systems from EMP attack. Although CAP is never mentioned, the commission found the cost difference between nonhardened and EMP-hardened communications equipment to be "a very small fraction" akin to one to three percent, further acknowledging how the cost for improved security "is modest by any standard—and extremely so in relation to both the war on terror and the value of the national infrastructures involved." William R. Graham et al., *Report of the Commission to Assess the Threat to the United States from Electromagnetic Pulse (EMP) Attack, Vol 1: Executive Report* (2004), http://www.empcommission.org/docs/empc_exec_rpt.pdf.

49. 1st Air Force Public Affairs, "1stAF's auxiliary capabilities surpass 10,000 pandemic man-days," CONR-1AF (AFNORTH), 20 May 2020, https://www.1af.acc.af.mil/News/Article-Display/Article/2192515/1stafs-auxiliary-capabilities-surpass-10000-pandemic-man-days/.

Appendix A

Activation of Civil Air Patrol (CAP) Coastal Patrols

Table A.1. Numbers, locations, and activation dates of coastal patrol bases and their start of operations

Base no.	Location	Activation date (all in 1942)	Start of patrol operations (all in 1942)
1	Bader Field/Atlantic City Municipal Airport, Atlantic City, NJ	28 Feb.	10 Mar.
2	Rehoboth Airport, Rehoboth, DE	28 Feb.	5 Mar.
3	Palm Beach County Park Airport, Lantana, FL[1]	30 Mar.	2 Apr.
4	Parksley Airport, Parksley, VA	16 Apr.	17 May
5	Flagler Beach Municipal Airport, Flagler Beach, FL[2]	11 Mar.	19 May
6	McKinnon St. Simons Island Airport, St. Simons Island, GA	12 May	20 May
7	Chapman Field, Miami, FL[3]	13 May	14 May
8	James Island Airport, Charleston, SC	23 May	30 May
9	Grand Isle Airport, Grand Isle, LA[4]	25 June	6 July
10	Beaumont Municipal Airport, Beaumont, TX	24 June	7 July
11	Raby Field/Pascagoula/Jackson County Airport, Pascagoula, MS	24 June	7 July
12	San Benito Municipal Airport, San Benito, TX[5]	8 July	24 July
13	Lowe Field/Sarasota Municipal Airport, Sarasota, FL[6]	9 July	7 Aug.
14	Atkinson Field/Panama City Airport, Panama City, FL	16 July	8 Aug.
15	Cliff Maus Municipal Airport, Corpus Christi, TX	20 July	7 Aug.
16	Dare County Regional Airport, Manteo, NC[7]	21 July	10 Aug.
17	Francis S. Gabreski Airport, Riverhead, NY	6 Aug.	18 Aug.
18	Falmouth/Coonamessett Airport, Falmouth, MA	25 Aug.	14 Sept.

Table A.1. (*continued*)

Base no.	Location	Activation date (all in 1942)	Start of patrol operations (all in 1942)
19	Portland International Jetport, Portland, ME	18 Aug.	1 Sept.
20	Hancock County-Bar Harbor Airport, Bar Harbor, ME	22 Aug.	5 Sept.
21	Michael J. Smith Field, Beaufort, NC	7 Sept.	27 Sept.

Sources: Office of Civilian Defense (OCD), Civil Air Patrol National Headquarters (CAPNHQ), Operations Orders No. 1, Activation of CAP Coastal Patrols, 30 November 1942, Folder 2, Box 6, Earle L. Johnson Papers (ELJ), Western Reserve Historical Society (WRHS); OCD, CAPNHQ, "Radio Stations Controlling CAP Coastal Patrol and Liaison Patrol Operations," 30 November 1942, Reel 44599; and CAPNHQ, untitled document marked "Confidential" listing movement of Coastal Patrol Bases, 15 February 1943, Reel 38920, Air Force Historical Research Agency (AFHRA).

Notes

1. On 19 March 1942, operations moved from Morrison Field, West Palm Beach, to Lantana Airport.

2. On 30 October 1942, operations moved from Daytona Beach Airport to Flagler Beach Municipal Airport.

3. Between 6-12 August 1942, operations moved from Miami Municipal Airport to Chapman Field.

4. On 1 August 1942, operations moved from New Orleans Airport to Grand Isle Airport.

5. On 28 December 1942, operations moved from Brownsville Airport to San Benito Municipal Airport.

6. On 9 October 1942, operations moved from Peter O. Knight Airport, Tampa, to Sarasota Municipal Airport.

7. On 27 October 1942, operations moved from Skyco Field/Manteo Municipal Airport to Naval Auxiliary Air Station Manteo.

Appendix B

CAP Coastal Patrol Personnel Killed on Active Duty

Table B.1. Names, hometowns, bases, and dates of those killed during their coastal patrol tenure

Rank/Name	Hometown	Base	Date
1st Lt Charles E. Shelfus*	Columbus, OH	2	21 July 1942
1st Lt Charles W. Andrews	Springfield, OH	14	30 Oct. 1942
1st Lt Lester E. Milkey	Sandusky, OH	14	30 Oct. 1942
1st Lt Alfred H. Koym	Rosenberg, TX	10	11 Nov. 1942
1st Lt James C. Taylor	Baton Rouge, LA	10	11 Nov. 1942
1st Lt Guy T. Cherry Jr.	Kinston, NC	21	16 Nov. 1942
1st Lt John H. Dean	Fort Worth, TX	10	16 Nov. 1942
1st Lt Robert D. Ward	Dallas, TX	10	16 Nov. 1942
1st Lt Frank M. Cook	Concord, NC	16	21 Dec. 1942
1st Lt Julian L. Cooper*	Nashville, NC	16	21 Dec. 1942
1st Lt Curtis P. Black*	North Olmsted, OH	14	4 Jan. 1943

Table B.1 *(continued)*

Rank/Name	Hometown	Base	Date
1st Lt Alvie T. Vaughen	Galion, OH	14	4 Jan. 1943
1st Lt Welles L. Bishop	Meriden, CT	20	2 Feb. 1943
1st Lt William B. Hites	Jamestown, NY	20	2 Feb. 1943
2nd Lt Drew L. King	Spartanburg, SC	8	9 Feb. 1943
2nd Lt Clarence L. Rawls	Charleston, SC	8	9 Feb. 1943
2nd Lt Martin E. Coughlin*	Kansas City, MO	11	26 Feb. 1943
1st Lt Paul W. Davis*	St. Louis, MO	11	26 Feb. 1943
1st Lt Harold O. Swift	Stanton, DE	2	6 Mar. 1943
1st Lt Delmont B. Garrett	Media, PA	2	19 Mar. 1943
2nd Lt Paul D. Towne	Peoria, IL	2	19 Mar. 1943
2nd Lt Donald C. Ferner	Tulsa, OK	14	3 Apr. 1943
1st Lt Gerald G. Owen	West Farmington, OH	14	3 Apr. 1943
1st Lt Ben Berger	Denver, CO	1	25 Apr. 1943
Capt Harry L. Lundquist	Gastonia, NC	21	27 June 1943
Flt Officer David S. Williams	Wallace, NC	21	27 June 1943

Source: Neprud, *Flying Minute Men*, 120–21; and Hudson, *Civil Air Patrol Fatalities*, 3.
*Asterisk indicates CAP member's body lost at sea

Appendix C
CAP Coastal Patrol Base Commanders March 1942–August 1943

Table C.1. Coastal patrol base commanders by base number and location

Base no.	Locations	Commanders
1	Atlantic City, NJ	Maj Gill Robb Wilson; Maj Wynant C. Farr
2	Rehoboth, DE	Maj Holger Hoiriis; Maj Hugh R. Sharp Jr.
3	West Palm Beach/Lantana, FL	Jacob M. Boyd; Maj Wright Vermilya Jr.
4	Parksley, VA	Maj Isaac W. Burnham II
5	Daytona Beach/Flagler Beach, FL	Maj Julius L. Gresham
6	St. Simons Island, GA	Inman Brandon; Maj Thomas H. Daniel Jr.
7	Miami, FL	Capt Van H. Burgin; Maj Lloyd H. Fales
8	Charleston, SC	1st Lt Cornelius O. Thompson; Maj Sidney B. Mahaffey; Maj Jack K. Moore
9	New Orleans/Grand Isle, LA	Maj Byron A. Armstrong; Maj Melvin A. Smith
10	Beaumont, TX	1st Lt William M. Cason; Maj George E. Haddaway
11	Pascagoula, MS	Maj Esmond Avery
12	Brownsville/San Benito, TX	Maj Benjamin S. McGlashan
13	Tampa/Sarasota, FL	Maj Peter J. Sones
14	Panama City, FL	Maj Robert E. Dodge; Maj Ernest T. Dwyer
15	Corpus Christi, TX	Maj William G. Green
16	Manteo, NC	Capt James L. Hamilton; Maj Allen H. Watkins

Table C.1 (*continued*)

Base no.	Locations	Commanders
17	Riverhead, NY	Maj Ralph Earle
18	Falmouth, MA	Capt J. Gordon Gibbs; Maj Ralph Earle
19	Portland, ME	Maj Milton V. Smith
20	Bar Harbor, ME	Maj James B. King
21	Beaufort, NC	Maj Frank E. Dawson

Source: CAPNHQ, untitled document marked "Confidential" listing CAP Coastal Patrol and Liaison Patrol Base Commanders, 15 February 1943, Reel 38920, AFHRA.

Appendix D

CAP Coastal Patrol Flight Hours
October 1942–August 1943

Table D. 1. Hours flown by coastal patrols, by month

Month	Escort patrol hours	Reconnaissance patrol hours	Total hours
Oct. '42	5,956:59	17,655:28	23,612:27
Nov. '42	6,619:49	14,936:21	21,556:09
Dec. '42	5,630:24	9,762:16	15,392:40
Jan. '43	2,994:02	8,930:22	11,924:24
Feb. '43	1,815:40	9,263:40	11,079:20
Mar. '43	2,275:00	9,382:30	11,657:30
Apr. '43	3,563:35	10,892:15	14,455:50
May '43	5,365:12	13,448:24	18,813:36
June '43	2,810:00	10,844:00	13,654:00
July '43	3,006:00	11,004:00	14,010:00
Aug. '43	1,816:00	10,321:00	12,182:00
Totals	41,852:41	126,440:16	168,292:57

Source: Table, "Hours Flown by Civil Air Patrol," compiled from Army Air Forces Antisubmarine Command monthly intelligence reports from October 1942 to August 1943, in Antisubmarine Command Historical Section, "CAP History of Operations (First Narrative)," 12 October 1943, Reel A4057, AFHRA.

Appendix E

Aircraft on Active CAP
Coastal Patrol Duty as of 28 April 1943

Aircraft on Active CAP Coastal

Base	Aeronca	Beechcraft	Bellanca	Buhl	Cessna	Cub Cruiser	Curtis	Fairchild	Fleet
1. Atlantic City, NJ	0	0	1	0	0	0	0	6	0
2. Rehoboth Beach, DE	0	0	0	0	0	0	0	13	0
3. Lantana, FL	0	0	0	0	1	0	0	2	0
4. Parksley, VA	0	1	0	0	1	0	0	5	1
5. Flagler Beach, FL	0	0	0	0	0	0	0	3	0
6. St. Simons Island, GA	0	0	0	0	1	1	0	7	0
7. Miami, FL	0	0	0	0	0	0	0	8	0
8. Charleston, SC	0	0	0	0	0	0	0	1	0
9. Grand Isle, LA	0	0	0	0	1	0	0	3	0
10. Beaumont, TX	0	0	0	0	1	0	0	6	0
11. Pascagoula, MS	0	0	0	0	1	0	0	3	0
12. San Benito, TX	0	1	0	0	0	0	0	8	0
13. Sarasota, FL	1	0	1	0	0	0	0	3	0
14. Panama City, FL	0	0	0	1	6	0	0	3	0
15. Corpus Christi, TX	0	2	0	0	2	0	0	5	0
16. Manteo, NC	0	0	0	0	0	0	2	1	0
17. Riverhead, NY	0	0	0	0	0	0	0	5	0
18. Falmouth, MA	0	0	0	0	0	0	0	6	0
19. Portland, ME	0	0	1	0	1	0	0	2	0
20. Bar Harbor, ME	0	0	0	0	0	0	0	3	0
21. Beaufort, NC	0	0	0	0	0	0	0	7	0
Aircraft Model Totals	1	4	3	1	15	1	2	100	1

Source: OCD, CAPNHQ, "Aircraft on Active CAP Coastal Patrol Duty by Type of Place," 28 April 1943, Folder "Submarine-

APPENDIX E | 195

Patrol Duty as of 28 April 1943

Howard	Luscombe	Monocoupe	Piper Cub	Rearwin	Ryan	Sikorsky	Stinson	Taylorcraft	Waco	Totals
0	0	0	0	0	0	0	14	0	11	34
0	0	0	0	0	0	1	3	0	1	18
0	0	0	0	0	0	1	20	0	0	24
0	0	0	0	0	1	0	9	0	4	22
0	0	3	0	0	0	0	8	0	4	18
0	0	0	0	0	2	0	8	0	5	24
0	0	0	0	1	0	0	10	0	0	19
0	0	0	0	0	0	0	14	0	0	15
0	0	0	0	0	0	0	9	0	7	21
0	0	0	0	1	0	0	14	0	2	24
0	1	1	0	0	0	0	8	0	3	17
0	0	0	0	0	0	1	8	0	3	21
0	0	0	0	0	0	0	8	0	4	17
0	0	0	0	0	0	0	5	0	2	17
1	1	0	0	1	0	1	8	0	5	26
0	0	0	0	1	0	0	9	0	2	15
0	0	0	0	0	0	1	9	0	6	21
0	1	0	0	0	0	1	10	0	0	18
0	0	0	0	1	0	0	7	1	5	18
0	1	0	0	0	0	0	9	1	2	16
0	0	0	1	0	0	0	7	0	3	18
1	4	4	1	5	3	6	197	2	69	423

on," Box 4, Entry 117, Office, Special Consultant to the Secretary of War, RG107, NARA.

Appendix F

"Definitely Damaged or Destroyed"

Reexamining CAP's Wartime Claims

In a 28 December 1943 restricted "Report of the Civil Air Patrol" to the Assistant Chief of Air Staff, Operations, Commitments and Requirements, CAP National Headquarters included a detailed summary about the coastal patrol operation that ran from 5 March 1942 to 31 August 1943. Among the figures listed are two highlighting the military nature of these civilian-flown missions. Namely, a report of 82 "bombs dropped against enemy submarines" and a claim of two "enemy submarines definitely damaged or destroyed."[1] In February 1944, the Navy published the August 1943 War Diary for the Eastern Sea Frontier, which also included the cumulative CAP coastal patrol statistics. The Navy war diary prefaced the information by noting that "the CAP Coastal Patrol left an interesting record of service."[2]

Since the fall of 1943, CAP has believed that its 18-month-long coastal patrol operation definitely damaged or destroyed two German U-boats. Following the conclusion of the war, this claim evolved within the organization to become a claim of destroying two enemy submarines, albeit with only circumstantial supporting evidence. Nevertheless, articles and press releases from CAP or the US Air Force as well as other accounts of CAP's coastal patrol effort repeat the claims of destroying submarines.[3] Over the course of researching CAP's coastal patrol history, the author published an examination of this wartime claim. This appendix is a revised, abridged version of that article.[4]

The majority of documented CAP coastal patrol submarine attacks date from May to November 1942. Throughout this period, approximately 42 U-boats patrolled at varying points along the East and Gulf coasts, during which time CAP reported 39 attacks on enemy submarines.[5] CAP's two incidents claiming to damage or destroy a submarine both occurred a day apart in July. The first incident occurred on 10 July 1942 approximately 14 miles off Cape Canaveral at position 28.43N, 80.30W. Aircraft from the Fifth Task Force, Daytona Beach, Florida, had only just begun armed patrols on 1 July with racks and simple bombsights installed by Army mechanics at Orlando

Army Air Base.[6] All base aircraft carried AN-M30 bombs, either as singles or as a pair.[7]

Details of the incident are fragmentary at best, but according to CAP and US Tenth Fleet records, one or possibly two CAP coastal patrol aircraft dropped three bombs on a reported submerged submarine at 1314 hours. The incident is not mentioned in the Fifth Task Force yearbook, but Tenth Fleet gave the incident two record numbers and the Joint Army-Navy Assessment Committee evaluated the results as "H" (insufficient evidence of presence of submarine) and later "J" (insufficient information to access or inconclusive).[8] That same July day, a Type VIIC submarine, *U-134*, commanded by *Kapitänleutnant* Rudolf Schendel, was sitting on the ocean floor, 26 miles from Cape Canaveral, but he reported no attacks nor sounds of explosions in his *Kriegstagebüch* (KTB) or war diary.[9]

The second incident forming CAP's damaged or destroyed claim has more substantial supporting evidence. Coincidentally, it occurred the day after the incident off Florida. Unlike the Fifth Task Force, Atlantic City's planes sported an array of bomb racks installed at Mitchel Field to carry the smaller 100-pound demolition bombs as well as the more formidable AN-M57 or Mk 17 bombs. None of the base aircraft had bombsights.[10] On 11 July 1942, one of the morning patrol aircraft from the First Task Force, Atlantic City, reported spotting a U-boat cruising on the surface off the coast of Absecon, New Jersey. After the reporting patrol returned to base, a Grumman G-44 Widgeon seaplane flown by Maj Wynant G. Farr and Capt John B. Haggin flew to the reported position and began a search for the submarine. Locating a faint oil slick, the men tracked its origin and concluded that the submerged submarine was moving parallel to shore. After patrolling for several hours over the location of the target, the men reported the submarine rose to periscope depth, at which point they dropped the Widgeon's two Mk 17 bombs, producing a spreading oil slick and bringing fragments of wood to the surface. Farr believed he saw the bow of the submarine break the surface of the water before sinking below.[11]

The Eastern Sea Frontier war diary entry for 11 July 1942 reports CAP sighting a submerged submarine at 39.07N, 74.13W, on course 280°, later revised to 39.15N, 74.13W, with "globs of oil appearing at distances of 15 feet and spreading." The entry notes that the latter position was 3 miles west of the wreck of the cargo ship *San Jose*, sunk after a collision on 17 January 1942. There is no mention of CAP attacking the object, but a Navy blimp, OS2U Kingfisher aircraft, patrol boats, and several coast guard cutters depth charged other positions in the area, bringing up wood and oil on the same day.[12] The Tenth Fleet assigned the attack incident no. 1083, occurring at 1545 hours, with a Joint Army-Navy Assessment Committee evaluation of "J."[13]

As with the incident of 10 July 1942, German war records show that a Type VIIC submarine, *U-89*, was patrolling slowly on a south/southwesterly course within 60 nautical miles of the shore. On the eleventh, the boat's commander, *Kapitänleutnant* Dietrich Lohmann, did not report any aircraft sightings much less attacks in his KTB, with the boat approximately 53 miles from the reported position of the CAP attack. Two days later, *U-89* was spotted and attacked by an aircraft approximately 50 miles east of Rehoboth Beach, Delaware, with three bombs causing slight damage to the submarine. CAP did not report this attack, which German researcher Axel Niestlé credits to a B-18 bomber of the Army Air Forces' 2nd Bomb Group.[14]

The reports of 10 and 11 July 1942 from the First and Fifth Task Forces arrived at CAP National Headquarters in short order for compilation with reports of the other task forces. Col Harry H. Blee, CAP National Headquarters' operations officer, oversaw the CAP coastal patrol effort during the war and received weekly reports from the task forces detailing total missions and hours flown, submarine sightings and/or attacks, irregularities at sea, floating bodies, or mines. Blee in turn submitted a weekly report tabulating the weekly figures for CAP's national commander, Maj Earle L. Johnson. Blee had no method to check the accuracy of the data within the weekly reports, relying entirely on the word of the respective task force commander.[15]

In his report to Johnson of 16 July 1942 covering the period of 9 to 14 July 1942, inclusive, Blee reported, "Civil Air Patrol planes dropped a total of seven bombs against enemy submarines. These bombing attacks resulted in the definite destruction of one submarine and the apparent damaging of another."[16] This assessment of damage or destruction appears to originate from Blee's analysis of the daily S-3 operations reports from the First and Fifth Task Forces, evidently independent from the assessments of Tenth Fleet, confirmed postwar by the surviving records of the German U-boat force.[17] A following report, issued months later by CAP National Headquarters on 20 October 1942, details that from 25 June to 29 July 1942, CAP coastal patrol aircraft "definitely damaged" two enemy craft.[18] By April 1943, prior to CAP's transfer from the Office of Civilian Defense to the War Department, a report authored by Capt Kendall K. Hoyt, CAP National Headquarters' intelligence officer, stated "2 enemy submarines have been destroyed or damaged by bombs from CAP planes."[19] This claim of two submarines damaged or destroyed subsequently found its way into the draft of the biennial report of the Army Air Forces.[20]

At the conclusion of the coastal patrol service on 31 August 1943, CAP tabulated its data. In August and September 1943, the Bureau of Public Relations for the War Department received data on coastal patrol operations "through channels" as reported by CAP National Headquarters.[21] The War Department released this CAP information in a press statement about the Antisubmarine Command on 10 December 1943, and CAP National Headquarters released its own version of this release, approved by the War Department's Bureau of Public Relations, one week later.[22] This official CAP statement of 17 December 1943 listed 173 submarines spotted, with 57 attacked with bombs or depth charges, and noted that CAP was "officially credited with sinking or damaging at least 2 [submarines], in addition to those sunk by Army or Navy aircraft called for the kill by CAP."[23] A restricted "Report of the Civil Air Patrol" published weeks later on 28 December 1943 by CAP National Headquarters for the Assistant Chief of Air Staff, Operations, Commitments and Requirements, included a summary of CAP coastal patrol operations postdated 3 September 1943. This statistical summary reported 82 "bombs dropped against enemy submarines" and listed two "enemy submarines definitely damaged or destroyed."[24] The only record or source that corroborates official credit appears to be Blee's July 1942 assessment of reports from the two CAP task forces.

In March 1944, the Army Air Forces Air Inspector released his report of an investigation of CAP from January to February 1944. Among the facts in the report, the document includes the September 1943 coastal patrol summary data "reported by the Civil Air Patrol," further reproduced by the Navy in the February 1944 war diary. The investigator wrote,

> Because of the conclusion of these operations, no detailed study of the accuracy of these claims was made. However, access was had to the evaluations given by the Navy to all claims of sinking submarines and it was determined therefrom [sic] that in the case of four claims made by the Civil Air Patrol, one was evaluated "No damage"; two, "Insufficient evidence of presence of submarine"; and a fourth, "Insufficient evidence of damage."

The armament carried by CAP planes during these operations was 100-pound demolition bombs. The question is presented as to how much damage a bomb of that weight and character could inflict upon a submarine under most favorable circumstances.[25]

The report raised clear doubts about the credibility of the CAP claims. On 31 August 1944, Johnson sent a reply detailing assorted corrections in response to the Air Inspector's report. Johnson does not mention, question, or rebuke the inspector's statements regarding the coastal patrol summary data.[26] In June 1945 when CAP National Headquarters submitted a historical report for the official history of the Office of Civilian Defense, the history noted CAP as "officially credited with sinking or damaging at least two [enemy submarines] in addition to those destroyed by planes or ships summoned by CAP."[27]

After the fall of the Third Reich, the records of the *Kriegsmarine*, notably those of the U-boat arm, were captured by the Allied forces. Analyzed in conjunction with the Ultra intercepts (decrypted German radio traffic), the Joint Army-Navy Assessment Committee was able to account for the fate of all of Germany's 1,167 U-boats. Of the 14 submarines confirmed sunk off the American Eastern and Gulf seaboards from March 1942 to August 1943, none were confirmed sunk by CAP; in fact the committee did not assign CAP credit for any U-boats.[28] The question of CAP damaging U-boats was not studied, but of those CAP attacked to receive Tenth Fleet incident numbers, the most promising evaluation recorded is "F," for "insufficient evidence of damage."[29]

Furthermore, CAP has no viable claim about the enemy submarines destroyed during its 18 months of patrol operations. Of the 14 submarines destroyed, American military forces, supported by physical or documentary evidence, received credit for definitively destroying 11 of these boats.[30] The *Kriegsmarine* never reported any submarine missing sent to American waters over the same period, and contemporary studies of all available data on the fate of the 1,167 U-boats corroborate the German record.[31]

From an examination of the existing archival evidence from Army, Navy, and German sources pertaining to CAP's coastal patrol effort, several conclusions are reached. CAP aircraft neither destroyed nor damaged any enemy submarines from 5 March 1942 to 31 August 1943. The claim by CAP of damaging or destroying enemy submarines appears to originate from within CAP's own national headquarters based on reports from the organization's coastal patrol task forces. The US military did not formally credit CAP with the destruction or damage of two enemy submarines, either during or after the conclusion of World War II.

Notes

1. CAPNHQ, Report of Civil Air Patrol, 28 December 1943, Appendix D, "Summary of CAP Coastal Patrol Operations," 3 September 1943, Folder 4, Box 1, ELJ, WRHS.
2. War Diary, Eastern Sea Frontier, August 1943, 40, NARA (via Fold3). The US Air Force Historical Study about civilian volunteer activities during World War II lists the same figures in the 28 December 1943 report, but attributes them to a report from December 31. Link, *Civilian Volunteer Activities*, 82–83.
3. "Serving, Saving, Shaping: Maj. Gen. Mark Smith Guides CAP Mission for State, Nation and Air Force," *AIRMAN Magazine*, 22 January 2019, http://airman.dodlive.mil/2019/01/22/serving-saving-shaping/.
4. Blazich, "'Definitely Damaged or Destroyed': Reexamining Civil Air Patrol's Wartime Claims," 19–30.
5. The figures for patrolling U-boats are taken from a review of KTBs with the assistance of Jerry Mason (Captain, USN, retired) and his website, http://www.uboatarchive.net. The figure for CAP attacks is both a review of information from the Eastern and Gulf Sea Frontier War Diaries and the records of the US Tenth Fleet's Antisubmarine Warfare (ASW) Analysis and Statistics Section files in Record Group 38 held in the National Archives. For the latter records, from 22 May to 5 November 1942, there are 32 incidents listed, albeit with two incident numbers for the same CAP attack. The Navy's Eastern and Gulf Sea Frontier War Diaries list additional CAP attacks, bringing the total to 39 by 5 November 1942. Between 5 November 1942 to 31 August 1943, CAP records only list an additional 6 attacks on enemy submarines, for a wartime total of 45.
6. Nanney, *CAP Coastal Patrol Base No. 5*, 12–15. The Fifth Task Force commenced operations at Daytona Beach Airport with its first patrols operating on 19 May 1942. In October, the base relocated to Flagler Beach, completing the move on 28 October. CAP 1st Lt John R. Tamm described the bombsight as "very simple–two paper clips attached 'just so' to the fuselage. We practiced at about a thousand feet until we could hit a target by lining up the target with the two clips. The pilot kept his right hand on his observer's knee and when he squeezed, his observer released the bomb." Keefer, *From Maine to Mexico*, 122.
7. Julius L. Gresham to Harry H. Blee, memorandum, subject: Report on Special Equipment and Supplies Furnished CAP Coastal Patrol #5, 11 November 1942, Reel 44599, AFHRA. Gresham noted 23 installed bombsights and 27 total bomb racks. Of the base's 23 aircraft, four had a brace of bombs, while 19 were armed with single bombs. Julius L. Gresham to Earle Johnson, memorandum, subject: Airplanes Stationed at Our Base, 13 July 1942, Reel 38919, AFHRA.
8. "Attack Record by Date," entries for 10–11 July 1942, incident no. 1062 and 1072, Folder "Chronological 1 May 42–30 June 42," Box 204, Tenth Fleet, ASW Analysis & Stat. Section, Series VIII: Assessments of Probable Damage Inflicted in Specific Anti-Submarine Warfare Incidents 1941–1945, Preliminary Evaluations–Attacks on U/B Assessments No. 1 1943; Incident No. 1062 and 1072, Box 72, Tenth Fleet, ASW Analysis & Stat. Section, Series VI: Reports and Assessment of Individual Anti-Submarine Warfare Incidents, 1941–1945, RG38, NARA; and Harry H. Blee to Lawrance Thompson, memorandum, subject: History–CAP Coastal Patrol Opera-

tions, 18 February 1944, Reel 38920, AFHRA. The Tenth Fleet lists incident 1062 on 10 July 1942 and incident 1072 on 11 July 1942, but the incident details are practically identical, notably the same location, time, and distance from land, leading the author to conclude these are one and the same incident.

9. KTB for Sixth War Patrol of *U-134*, entry for 10 July 1942, http://uboatarchive.net/U-134/KTB134-6.htm.

10. Wynant C. Farr to Harry H. Blee, memorandum, subject: Report on Special Equipment and Supplies, 9 November 1942, Reel 44599, AFHRA.

11. Neprud, *Flying Minute Men*, 18–19; and Weidenfeld, "Search for the Haggin-Farr Sub Kill," 5–8. Farr wrote a letter on 12 July 1942 of the incident to Johnson, who in turn passed the letter to Lieutenant General Arnold, who read the report aloud in an off-the-record session of a Senate Appropriations Committee. Earle L. Johnson to Wynant C. Farr and John B. Haggin, 17 July 1942, provided by Lt Col Gregory Weidenfeld.

12. War Diary, Eastern Sea Frontier, 11 July 1942, 1307 EWT (Eastern War Time), NARA (via Fold3).

13. "Attack Record by Date," entry for 12 July 1942, incident no. 1083, Folder "Chronological 1 May 42–30 June 42," Box 204, Tenth Fleet, ASW Analysis & Stat. Section, Series VIII: Assessments of Probable Damage Inflicted in Specific Anti-Submarine Warfare Incidents 1941–1945, Preliminary Evaluations–Attacks on U/B Assessments No. 1 1943; Incident No. 1083, Box 72, Tenth Fleet, ASW Analysis & Stat. Section, Series VI: Reports and Assessment of Individual Anti-Submarine Warfare Incidents, 1941–1945, RG38, NARA.

14. KTB for Second War Patrol of *U-89*, entries for 11 and 13 July 1942, http://uboatarchive.net/KTB89-2.htm; database of air attacks off US east coast reported by German U-boats from 1942–1943, provided to author by German U-boat historian Dr. Axel Niestlé.

15. War Department, Headquarters of the Army Air Forces, Dudley M. Outcalt to Air Inspector, memorandum, subject: Survey of the Civil Air Patrol, 8 March 1944, 6, Folder 3, Box 5, ELJ, WRHS.

16. Harry H. Blee to Earle L. Johnson, memorandum, subject: Coastal Patrol Operations, 9 to 15 July 1942, Inclusive, 16 July 1942, Folder "Civil Air Patrol," Box 7, Entry 1, Miscellaneous Records, Director's Office, Feb. 1942–June 1944, CAP–Budget Estimates, RG171, NARA.

17. Harry H. Blee to Lawrance Thompson, memorandum, subject: History–CAP Coastal Patrol Operations, 18 February 1944, Reel 38920, AFHRA.

18. OCD, CAPNHQ, Civil Air Patrol Coastal Patrol Operations, 20 October 1942, Reel 38919, AFHRA.

19. OCD, CAPNHQ, Kendall K. Hoyt, Special Report of the Civil Air Patrol, April 1943, 7, Reel A2992, AFHRA. In a second, "Brief Report on Civil Air Patrol" of April 1943, Hoyt reported CAP being "credited with destroying or damaging two" submarines. OCD, CAP, Kendall K. Hoyt, Brief Report on Civil Air Patrol, April 1943, Reel A2992, AFHRA.

20. Assistant Chief of the Air Staff, Intelligence, Historical Division, "Draft for Biennial Report of the Army Air Forces, 1 July 1941 to 30 June 1943," 86, Folder "Military Report General May 1943," Box 183, Henry Harley Arnold Papers (HHA), Library of Congress (LOC).

21. Kendall K. Hoyt to Historical Division, Assistant Chief of Air Staff, Intelligence, memorandum, subject: Civil Air Patrol, Week ended 20 November 1943, 20 November 1943, Folder 4, Box 1, ELJ, WRHS.

22. Kendall K. Hoyt to Historical Division, AC/AS, Intelligence, memorandum, subject: Civil Air Patrol, Week ended 18 December 1943, 18 December 1943, Folder 4, Box 1, ELJ, WRHS.

23. CAPNHQ, *CAP Bulletin* 2, no. 51 (17 December 1943): 1–2.

24. CAPNHQ, Report of Civil Air Patrol, 28 December 1943, Appendix D, "Summary of CAP Coastal Patrol Operations," 3 September 1943, Folder 4, Box 1, ELJ, WRHS.

25. War Department, Headquarters of the Army Air Forces, Dudley M. Outcalt to Air Inspector, memorandum, subject: Survey of the Civil Air Patrol, 8 March 1944, 30–31, Folder 3, Box 5, ELJ, WRHS.

26. Earle L. Johnson to Commanding General, Army Air Forces, memorandum, subject: Report of Air Inspector's Investigation of Civil Air Patrol, dated 8 March 1944, 31 August 1944, Folder 3, Box 5, ELJ, WRHS.

27. Earle L. Johnson to Elwyn A. Mauck, 21 June 1945, with attachment, "History of CAP," 25 June 1945, 11, Folder "Civil Air Patrol June 1–," Box 21, Office of Civilian Defense, National Headquarters, General Correspondence, 1941–May 1945, Civil Air Patrol, RG171, NARA.

28. Navy Department, Chief of Naval Operations, *German, Japanese and Italian Submarine Losses–World War II, (OPNAV-P33-100)* (Washington, DC: GPO, May 1946), 3–9. CAP-USAF was informed of this by November 1973 in response to an inquiry to the Department of the Navy's Naval Historical Center while trying to locate information on the two destruction claims. After finding no information to support or directly refute the matter, CAP-USAF decided to accept the story "as it has been told." Dean C. Allard to William T. Capers III, 27 November 1973; Frank O. Lowry to William T. Capers III, memorandum, subject: Sub sinkings, 12 December 1973; and William T. Capers III to Frank John E. Blake, 12 December 1973, Folder "INF34: Sub Sinking," CAP-NAHC.

29. "Attack Record by Date," entries for 2 June 1942, incident no. 777, Folder "Chronological 1 May 42–30 June 42," Box 204, Tenth Fleet, ASW Analysis & Stat. Section, Series VIII: Assessments of Probable Damage Inflicted in Specific Anti-Submarine Warfare Incidents 1941–1945, Preliminary Evaluations–Attacks on U/B Assessments No. 1 1943, RG38, NARA. The closest enemy submarine, the Type IXB submarine *U-106*, reported no attack on 2 June 1942 and was approximately 300 miles away from the reported time and position of the attack.

30. The 11 submarines sunk by the U.S. military include eight sunk by the Navy (*U-656, U-503, U-85, U-158, U-576, U-166, U-521, U-84*), two by the Coast Guard (*U-352, U-157*), and one by the Army Air Forces (*U-701*). Regarding the three other boats sunk off the U.S. East Coast, they are *U-215*, sunk by the British Royal Navy antisubmarine trawler HMS *Le Tiger*; *U-754*, sunk by a Royal Canadian Air Force Hudson bomber; and *U-176*, sunk by the Cuban Navy patrol boat *CS 13*. Niestlé, *German U-boat Losses*, 41, 56, 71, 78, 86, 114, 121–22, 124.

31. Niestlé, *German U-boat Losses*, 3. Combining American, British, Canadian, and German records, Axel Niestlé's work is the most exhaustive study of the fate of all of Germany's U-boats in World War II. He graciously shared his database of reported air attacks against U-boats off the East and Gulf Coasts of the United States from March 1942 until August 1943 with the author in the course of his research. No attacks, including those where the attacker remains unknown, match those reported by CAP in the Eastern or Gulf Sea Frontier war diaries.

＃ Appendix G

Proposed Plan for Organization of Civil Air Defense Resources

CONTENTS

Letter of transmittal.

Immediate objectives.

Ultimate objectives.

Principles of personnel.

Procedure for national organization.

Procedure for state organization.

Immediate strength objective table.

Federal scale of assistance for immediate objective.

June 24, 1941

The Honorable Fiorello La Guardia
Office of Civilian Defense
Washington, D. C.

Dear Mr. La Guardia:

We propose organization of the "Civil Air Defense Services". This preliminary report sets forth two objectives:

1. The immediate organization of the civil air resources available.

2. The ultimate civil development essential to any sound foundation for air power.

The immediate organization of the existing civil air resources is essential to national defense now. The ultimate development will be no less essential tomorrow.

Neither is this war nor will any future war be one of military establishments but of peoples. Civilian defense starts with the recognition of that fact. The people have skills and abilities. Recognizing their vulnerability in modern war they insist in using their several skills and abilities in self defense, for national defense is now self defense.

The personnel of civil aviation are patriotic, skillful and energetic. They know that if properly organized they can render a high degree of utility. Casual talk of grounding civil aviation in case of military action elicits bitter and just protest. Nevertheless such action might be necessary unless organization is effected and civil flying disciplined and turned to objective defense utility. This organization we recognize as one of the vital jobs of the Office of Civilian Defense. You have so recognized it.

- 2 -

The Honorable Fiorello La Guardia   June 24, 1941

    Air commerce and industry will spearhead human activity for the next quarter century. Nothing less than an all-out civil and military air program will serve the United States. We propose to organize what civil resources we now have and build a foundation upon which to create more. We appreciate the confidence you expressed in assigning us this task.

                          Respectfully,

                          Gill Robb Wilson, Chairman
                          Committee on Plans for
                          Civil Air Defense

IMMEDIATE OBJECTIVES TO BE ACCOMPLISHED BY FORMATION
OF THE CIVIL AIR DEFENSE SERVICES

I.

Development within civil aviation of a morale and discipline essential to the requirements of effective national defense.

II.

Organization of the accumulated knowledge and experience of the great civil pilot pool not eligible for combatant service but potentially eligible for auxiliary duty in emergency.

III.

A rigid survey of civil airmen which would classify the loyal and trustworthy and ascertain the subversive and untrustworthy. All aliens shall be grounded except by specific individual privilege.

IV.

Creation of a trained and disciplined civil air component to stand guard over the activities of the two thousand scattered landing fields and airports not occupied by the military or scheduled air transport operators.

V.

To provide a source capable of furnishing air transportation of personnel or material upon request of proper military or civil authority.

VI.

To provide a group capable of patrol against flight over restricted areas of industry, potable water supply, cities, arsenals or other areas where sabotage or the gathering of information from the air might interfere with national defense.

- 2 -

Immediate Objectives.

VII.

To provide an organized service for the marking out and guarding of areas which although not airports might be used for emergency landings in case of necessity.

VIII.

To provide an organized service available for use in emergency disaster when military aircraft and personnel were otherwise engaged.

IX.

To create an organization trained in the accurate observation and identification of all aircraft and with knowledge of aviation procedures, which knowledge would be available to all civil defense components.

X.

To create a service with aggregate knowledge of the entire terrain of the nation and capable of searching for aircraft which crashed or might be lost through misadventure.

XI.

To have available a service trained in observation of highway traffic and capable of action under evacuation or disaster conditions.

XII.

To create a service possessing full and accurate aviation information and charged with responsibility for conveying such information to the public school components of the nation.

XIII.

To create an organization capable of assistance to and cooperation with military personnel who might require service at an airport or emergency field not occupied by a military unit.

ULTIMATE OBJECTIVES OF THE CIVIL AIR DEFENSE SERVICES

500,000 civil pilots by 1945

1,500,000 ground service personnel by 1945

250,000 civil aircraft by 1945

6,000 airports and landing strips by 1945

METHODS FOR ATTAINMENT

Education

A. There are 232,174 public elementary schools in the United States. The federal government shall allot through the state boards of education funds to provide:

    Elementary text books on aviation subjects and history

    Materials for model building and model kits

    Achievement rewards for excellence in scholarship

B. There are 25,467 public high schools in the United States. The federal government shall allot through the state boards of education funds to provide:

    Advanced text books on aviation subjects

    Materials for power model building

    Materials for soaring plane construction in vocational work

    Laboratory materials for meteorological study

    Laboratory materials for engine and plane construction

    Achievement rewards for excellence (scholarships)

- 2 -

Ultimate Objectives.

C. There are in excess of 750,000 school teachers in the United States. The federal government shall make available to any teacher physically and otherwise qualified the sum of $250.00, payable when the teacher has taken flight training and attained a private pilot's license.

D. The federal government shall make available to any public school teacher, without cost, the course in ground instruction as provided by applicants for CPTP students.

E. The federal government shall establish one thousand national scholarships to be competed for annually by high school students in the field of general and applied aviation knowledge, such scholarships to consist of further study in institutions of higher learning.

F. There are now employed in various programs of the National Youth Administration 876,222 young men and women. It is our proposal to have the federal government make a survey of the job classifications in all components of aviation and upon the basis of that survey to establish under the National Youth Administration training classes to equip the members of the NYA for such jobs and professions.

Development

A. The training up to the status of a private pilot's license of 500,000 young men and women. The following steps are suggested:

1. Those who take CPTP training as preliminary to military service shall receive that training on the present basis except that the CPTP program shall be quadrupled for the next fiscal year, and available to all with a high school education.

2. Any young man or woman who shall on his own volition learn to fly and attain a private pilot's license and shall after so learning enroll in the Civil Air Defense Services shall receive a bonus of $250.00 upon enlistment.

- 3 -

UltimateObjectives.

B. The federal government shall through the RFC assist in financing up to sixty per cent of the purchase price of any new aircraft under $1,500 list price, when such aircraft shall be purchased by an individual or combination of not to exceed three individuals who shall enlist and volunteer the use of their aircraft in the service of the CADS.

C. The federal government shall utilize the vast number of unemployed wood workers and the unused productive floor space of the furniture industry to develop the gliding and soaring plane industry and shall encourage and assist the companies so engaged at present and shall make available at cost to any club of five qualified students such a plane when the club shall be sponsored by a public school teacher.

D. The federal government shall pay a bonus to the manufacturer of any aircraft with a list price of less than fifteen hundred dollars provided fifty per cent of the total employees of the corporation are female, and provided that the bonus shall not be in excess of twenty per cent of the cost of the aircraft to the manufacturer.

E. There are in the Civilian Conservation Corps employed at the present time 213,358 young men. It is proposed that the Civilian Conservation Corps be immediately turned to the building of landing strips on areas of land owned or nominally rented by the government; and that in the Civilian Conservation Corps aviation ground courses be given to all who seek the knowledge as an adjunct to their daily work and to make that work more understandable and interesting.

F. Women as instructors in ground and flight training have proven exceptional merit. It is proposed that secondary instruction be given to all women pilots who signify their willingness to engage in such instruction either in the employ of CPTP operators or other fixed base operators.

ULTIMATE OBJECTIVES OF THE CIVIL AIR DEFENSE SERVICES
WORLD LEADERSHIP FOR AMERICAN CIVIL AVIATION
BY 1 9 4 5

THE PYRAMID OF CIVIL AIR POWER

Immediate
organization
of available
civil air resources
into
"The Civil Air Defense Services"
Transfer of emphasis in government policy from regulation to promotion
Priorities for materials used in fabrication of light aircraft and soaring planes
Federal assistance in aircraft purchase financing
A greater proportion of women in aviation industry
Federal aid for a truly comprehensive air training program
Utilization of national youth service to develop facilities
Simplified regulations and segregation of diverse air traffic
Indoctrination of youth through the public schools of the United States
Organized use of press and radio for public indoctrination in aviation

PRINCIPLES UPON WHICH THE CIVIL AIR DEFENSE SERVICES SHALL BE ESTABLISHED.

I.

The services of all personnel shall be voluntary and unremunerated.

II.

The use of all equipment voluntarily used by the personnel of the Civil Air Defense Services shall be accepted without pay or reward except as specified in the schedule hereinafter found.

III.

No person shall be accepted for service in the Civil Air Defense Services until rigorous investigation has shown such person to be qualified for such service and worthy of the confidence of the national defense authorities.

IV.

Each person accepted for service in the Civil Air Defense Services shall take the oath of allegiance to the United States and a pledge to devote the specified time required by the training directive.

V.

The Civil Air Defense Services shall attempt to set such an example of voluntary devotion to duty and self-imposed discipline as shall create the highest possible morale throughout civil aviation in the United States.

PROCEDURE FOR ORGANIZATION OF THE CIVIL AIR DEFENSE SERVICES

I.

An Aviation Section shall be created in the Office of Civilian Defense.

II.

This Aviation Section shall create an organization from the available civil air resources of the United States.

III.

This organization shall be known as the "Civil Air Defense Services."

IV.

The structure of the Civil Air Defense Services shall conform to that of the United States Army Air Corps.

V.

Enrollment in the Civil Air Defense Services shall be on a voluntary basis.

VI.

The Civil Air Defense Services shall be organized through the State Defense Councils.

VII.

Each state shall comprise a WING of the Civil Air Defense Services.

VIII.

Each Wing shall operate by and under the authority of its State Defense Council.

IX.

The headquarters staff of the Civil Air Defense Services shall correlate the activities of the state Wings.

- 2 -

Procedure.

X.

The Aviation Section of the Office of Civilian Defense shall furnish to each State Defense Council:

    (a) A uniform plan for organization of the state Wing.

    (b) A uniform training program.

    (c) A uniform service objective.

    (d) Financial assistance based on a schedule hereinafter stated.

PROCEDURE FOR ORGANIZATION OF A STATE WING OF THE
CIVIL AIR DEFENSE SERVICES

I.

Each State Defense Council shall appoint an aviation committee of the Council and certify such committee to the Aviation Section of the Office of Civilian Defense.

II.

The Aviation Section of the Office of Civilian Defense shall provide the aviation committee of the State Defense Council with:

 (a) A uniform plan for organization of the state unit.

 (b) A uniform training program.

 (c) A uniform service objective.

 (d) A schedule of financial assistance.

III.

The aviation committee of the State Defense Council shall make a survey of the available civil air resources of the state and, in the light of the specifications furnished, shall draw up plans for organization or a state Wing of the Civil Air Defense Services such as will best conform to the civil air defense needs of the state and shall submit this plan to the State Defense Council for approval.

IV.

Upon approving the plan, the State Defense Council shall certify such approval to the Aviation Section of the Office of Civilian Defense.

V.

The Aviation Section of the Office of Civilian Defense shall review such plan and, upon approval, shall authorize the aviation committee of the State Defense Council to proceed with organization and shall certify to the Office of Civilian Defense that the State Defense Council is eligible to receive such financial aid as is called for in the schedule according to the approved plan.

IMMEDIATE STRENGTH OBJECTIVE OF THE CIVIL AIR
DEFENSE SERVICES

I.

To secure and organize on a voluntary basis the services of 5000 civil aircraft owners and their aircraft, classified as follows:

(a) 3600 land planes powered by engines of less than 100 horsepower

(b) 400 seaplanes powered by engines of less than 100 horsepower

(c) 500 landplanes powered by engines of more than 100 and less than 250 horsepower

(d) 100 seaplanes powered by engines of more than 100 and less than 250 horsepower

(e) 400 land and sea planes powered by engines of more than 250 horsepower

II.

To secure and organize on a voluntary basis the services of 10,000 civil pilots who do not own aircraft but who will serve as co-pilots with aircraft owners or will adjust their flight time on rented aircraft to the service of the CADS.

III.

To secure and organize on a voluntary basis the services of 20,000 mechanics, shop workers, airport personnel, radio experts and others who by reason of aviation experience or knowledge may render service to the CADS.

IV.

To secure and organize into national objective at least one teacher in every public school in the United States to promote and develop aviation education in that school.

- 2 -

Immediate Strength Objective.

V.

To bring about through the mediation of the National Aeronautic Association a uniform objective program for all organized groups in civil aviation.

VI.

To enlist the voluntary services of all fixed-base operators in making their equipment available at favorable rentals to qualified pilots of the CADS.

VII.

To enlist the services of the Aviation Writers Association in acquainting the American public with the work of the CADS.

VIII.

To secure priorities in aviation fuel and lubricants, repair parts and manufacturers' materials for aircraft whose owners or purchasers make affidavits that such equipment is to be used in the service of the CADS.

IX.

To enlist in every state and municipal police department the services of some member with aviation experience who can act as liaison officer between local defense authority and the local unit of the CADS.

X.

To enlist in the CADS all personnel of the Civil Aeronautics Administration and the aircraft operated by the same.

SCALE OF FEDERAL ASSISTANCE TO THE CADS

I.

While no salaries or rewards are to be paid to personnel of the CADS, the voluntary use of aircraft by owners might be practically impossible unless assistance in meeting costs of fuel, lubrication and maintenance is provided. The scale of assistance suggested is not designed to fully meet such costs but only to be sufficient to make the equipment available.

II.

While it shall be clearly understood and provided in the enlistment papers of the CADS that no claims for damages, loss of property or life shall be entertained against the federal or state governments because of service in the CADS, it is suggested that a form of insurance such as that available to CPTP students be made available to flying personnel of the CADS.

III.

It is suggested that to give federal status and recognition to the CADS all personnel shall be placed upon the federal payroll at the total wage of ONE DOLLAR.

IV.

Any expense for organization, quarters, equipment, clerical help or other costs incidental to a state Wing shall be provided by the State Defense Council which authorizes the Wing organization.

V.

The CADS shall provide for each member of the CADS enrolled through any source a suitable brevet or insignia emblematic of membership in the CADS in whatever type of activity the individual shall take service. Such emblems shall be limited in cost to less than one dollar per person.

VI.

For aircraft flown on official duty for the Civil Air Defense Services the schedule of assistance to aircraft owners shall be as follows:

(a) $3.00 per hour.................aircraft under 100 horsepower; not to exceed $150.00 to any owner in any one annual period. $3.00 per hour to any operator who makes available to pilot personnel of the CADS his aircraft of under 100 horsepower; not to exceed payments of $900.00 to any one operator in any one annual period.

(b) $5.00 per hour.................aircraft over 100 horsepower and under 250 horsepower; not to exceed $250.00 to any owner in any one annual period. $5.00 per hour to any operator who makes available to pilot personnel of the CADS his aircraft of this classification; not to exceed payments of $1,500.00 to any one operator in any one annual period.

(c) $7.50 per hour.................aircraft over 250 horsepower; not to exceed $375.00 to any owner in any one annual period. $7.50 per hour to any operator who makes available to pilot personnel of the CADS his aircraft of this classification; not to exceed $2,250.00 to any one operator in any one annual period.

Payment in all cases shall be made only when official time flown has been certified to the CADS by the State Defense Council and payment shall be made through the State Defense Council.

Bibliography

Archives

(Archive content is cited heavily in chapter endnotes. The sources listed here provide direct links to the archives noted.)

Air Force Historical Research Agency (AFHRA), Maxwell AFB, AL. https://www.afhra.af.mil/.

Civil Air Patrol National Archives and Historical Collections (CAP-NAHC), Col Louisa S. Morse Center for Civil Air Patrol History, Joint Base Anacostia-Bolling, Washington, DC. https://history.cap.gov/.

   Containing the August William Schell Collection (AWS), Barry L. Spink Collection (BLS), CAP Historical Foundation Collection (CHF), Coastal Patrol Base No. 16 Collection (CP16), Henry E. Phipps Papers, and the Kendall King Hoyt Papers (KKH).

Dare County Regional Airport Museum, Manteo, NC. https://www.darenc.com/departments/airport/museum.

Harry S. Truman Presidential Library and Museum, Independence, MO. https://www.trumanlibrary.gov/.

Manuscript Division, Library of Congress (LOC), Washington, DC. https://www.loc.gov/collections/?fa=partof:manuscript+division.
   Containing the Henry Harley Arnold Papers (HHA).

National Archives and Records Administration (NARA), College Park, MD. https://www.archives.gov/.

Naval History and Heritage Command, Washington Navy Yard, DC. https://www.history.navy.mil/.

Outer Banks History Center, Manteo, NC. https://archives.ncdcr.gov/researchers/outer-banks-history-center.

Southern Historical Collection, The Louis Round Wilson Library Special Collections, University of North Carolina at Chapel Hill, NC. https://library.unc.edu/wilson/shc/.

State Archives of North Carolina, Raleigh, NC. https://archives.ncdcr.gov/.

The Institute of Heraldry (TIOH), Fort Belvoir, VA. https://tioh.army.mil/.

Western Reserve Historical Society (WRHS), Cleveland, OH. https://www.wrhs.org/.
   Containing the Earle L. Johnson Papers (ELJ).

Books

19th Patrol Force. N.p., [1943–44?].

Baer, George W. *One Hundred Years of Sea Power: The U.S. Navy, 1890–1990.* Stanford, CA: Stanford University Press, 1994.

Blair, Clay. *Hitler's U-Boat War: The Hunted, 1942–1945.* New York: Random House, 1998. Blair, Clay. *Hitler's U-Boat War: The Hunted, 1942-1945.* New York: Random House, 1998. (Referred to in endnotes as *The Hunted.*)

———. *Hitler's U-Boat War: The Hunters, 1939–1942.* New York: Random House, 1996. (Referred to in endnotes as *U-Boat War.*)

Boudreau, James C. *CAPCP Base-17, 1942–1943.* New York: New Era Lithograph Co., 1944.

Burnham, Frank A. *Hero Next Door.* Fallbrook, CA: Aero Publishers Inc., 1974.

C.C.P. Base 10: Pass in Review! Beaumont, TX: [1943–44?].

Coates, E. J., ed. *The U-Boat Commander's Handbook.* [In German.] Translated by US Navy, 1943. Gettysburg, PA: Thomas Publications, 1989.

Colby, Charles B. *This Is Your Civil Air Patrol: The Purpose, Cadet Program, Equipment of the U.S. Air Force Auxiliary.* New York: Coward-McCann Inc., 1958.

Cremer, Peter. *U-Boat Commander: A Periscope View of the Battle of the Atlantic.* Translated by Lawrence Wilson. Annapolis, MD: Naval Institute Press, 1985.

Cressman, Robert J. *The Official Chronology of the U.S. Navy in World War II.* Annapolis, MD: Naval Institute Press, 2000.

Dallek, Matthew. *Defenseless Under the Night: The Roosevelt Years and the Origins of Homeland Security.* New York: Oxford University Press, 2016.

Daso, Dik Alan. *Hap Arnold and the Evolution of American Airpower.* Washington, DC: Smithsonian Institution Press, 2000.

Dimbley, Jonathan. *The Battle of the Atlantic: How the Allies Won the War.* New York: Oxford University Press, 2016.

Doenitz, Karl. *Memoirs: Ten Years and Twenty Days.* New York: World Publishing Company, 1959.

Finnegan, John P. *Against the Specter of a Dragon: The Campaign for American Military Preparedness, 1914–1917.* Westport, CT: Greenwood Press, 1974.

Frebert, George J. *Delaware Aviation History*. Dover, DE: Dover Litho Printing, Co., 1998.
Freitus, Joseph, and Anne Freitus. *Florida: The War Years, 1938–1945*. Niceville, FL: Wind Canyon Pub., Inc., 1998.
Gannon, Michael. *Black May: The Epic Story of the Allies' Defeat of the German U-Boats in May 1943*. New York: HarperCollins Pubs. Inc., 1998.
―――. *Operation Drumbeat: The Dramatic True Story of Germany's First U-Boat Attacks along the American Coast in World War II*. New York: Harper Perennial, 1991.
Gordon, Dennis. *The Lafayette Flying Corps: The American Volunteers in the French Air Service in World War I*. Atglen, PA: Schiffer Pubs. Ltd., 2000.
Harries, Meirion, and Susie Harries. *The Last Days of Innocence: America at War, 1917–1918*. New York: Random House, 1997.
Hickam, Homer H. *Torpedo Junction: U-Boat War Off America's East Coast, 1942*. Annapolis, MD: United States Naval Institute Press, 1996.
Hill, Jim Dan. *The Minute Man in Peace and War: A History of the National Guard*. Harrisburg, PA: Stackpole Co., 1964.
Hoyt, Edwin P. *U-Boats Offshore: When Hitler Struck America*. New York: Stein and Day, 1978.
Hyde, Charles K. *Arsenal of Democracy: The American Automobile Industry in World War II*. Detroit: Wayne State University Press, 2013.
Jeffers, H. Paul. *The Napoleon of New York: Mayor Fiorello La Guardia*. New York: John Wiley and Sons Inc., 2002.
Jordan, David M. *Robert A. Lovett and the Development of American Air Power*. Jefferson, NC: McFarland and Co., Inc., Pubs., 2019.
Keefer, Louis E. *From Maine to Mexico: With America's Private Pilots in the Fight against Nazi U-Boats*. Reston, VA: COTU Publishing, 1997.
Laslie, Brian D. *Architect of Air Power: General Laurence S. Kuter and the Birth of the US Air Force*. Lexington: University Press of Kentucky, 2017.
Love, Terry M. *L-Birds: American Combat Liaison Aircraft of World War II*. New Brighton, MN: Flying Books International, 2001.
Mason, David. *U-Boat: The Secret Menace*. New York: Ballantine Books Inc., 1968.

Mellor, William B., Jr. *Sank Same*. New York: Howell, Soskin Publishers, 1944.

Millett, Allan R., and Peter Maslowski. *For the Common Defense: A Military History of the United States of America*. Rev. ed. New York: The Free Press, 1994.

Mitchell, William. *Our Air Force: The Keystone to National Defense*. New York: D. P. Dutton and Co., 1921.

———. *Winged Defense: The Development and Possibilities of Modern Air Power—Economic and Military*. New York: G. P. Putnam's Sons, 1925.

Morison, Samuel Eliot, ed. *The Battle of the Atlantic, September 1939–May 1943*. Vol. 1 of *History of the United States Naval Operations in World War II*. Boston: Little, Brown and Company, 1947.

———. *The Two-Ocean War: A Short History of the United States Navy in The Second World War*. Boston: Little, Brown and Company, 1963.

Mosley, Zack. *Brave Coward Zack*. St. Petersburg, FL: Valkyrie Press Inc., 1976.

Mulligan, Timothy P. *Neither Sharks nor Wolves: The Men of Nazi Germany's U-Boat Arm, 1939–1945*. Annapolis, MD: Naval Institute Press, 1999.

Murray, Williamson, and Allan R. Millett. *A War to Be Won: Fighting the Second World War*. Cambridge, MA: The Belknap Press, 2000.

Neprud, Robert E., and Zach Mosley. *Flying Minute Men: The Story of the Civil Air Patrol*. Rev. ed. Washington, DC: Office of Air Force History, 1988.

Niestlé, Axel. *German U-Boat Losses During World War II: Details of Destruction*. London: Frontline Books, 2014.

Offley, Edward. *The Burning Shore: How Hitler's U-Boats Brought World War II to America*. New York: Basic Books, 2014.

Overy, Richard J. *Why the Allies Won*. New York: W. W. Norton and Co., 1996.

Pisano, Dominick A. *To Fill the Skies with Pilots: The Civilian Pilot Training Program, 1939–1946*. Washington, DC: Smithsonian Institution Press, 2001.

Ragsdale, Kenneth Baxter. *Wings Over the Mexican Border: Pioneer Military Aviation in the Big Bend*. Austin: University of Texas Press, 1984.

Ritchie, Donald A. *James M. Landis: Dean of the Regulators*. Cambridge, MA: Harvard University Press, 1980.

Roskill, Stephen W. *The War at Sea, 1939–1945.* 4 vols. 1954–1961. Reprint, Uckfield, UK: Naval and Military Press Ltd., 2004.

Runyan, Timothy J., and Jan M. Copes, eds. *To Die Gallantly: The Battle of the Atlantic.* Boulder, CO: Westview Press, 1994.

Schlegel, Marvin W. *Virginia on Guard: Civilian Defense and the State Militia in the Second World War.* Richmond, VA: Virginia State Library, 1949.

Schoenfeld, Max. *Stalking the U-Boat: USAAF Offensive Antisubmarine Operations in World War II.* Washington, DC: Smithsonian Institution, 1995.

Sloan, James J., Jr. *Wings of Honor: American Airmen in World War I.* Atglen, PA: Schiffer Pubs. Ltd., 1994.

Southern Flight. *Civil Air Patrol Handbook, 1944 Edition.* 2nd ed. Dallas, TX: Southern Flight, 1944.

Swanborough, Gordon, and Peter M. Bowers. *United States Military Aircraft Since 1909.* Washington, DC: Smithsonian Institution Press, 1989.

Ten Eyck, Andrew. *Jeeps in the Sky: The Story of Light Planes in War and Peace.* New York: Commonwealth Books Inc., 1946.

The Council of State Governments. *The Book of the States, 1941–1942.* Vol. 4. Chicago: The Council of State Governments, 1941.

Thompson, David Robinson, and Sandra Thompson. *Palm Beach: From the Other Side of the Lake.* New York: Vantage Press, 1992.

Topp, Erich. *The Odyssey of a U-Boat Commander: Recollections of Erich Topp.* Translated by Eric C. Rust. Westpoint, CT: Praeger Publishers, 1992.

Warner, Melvin J., and George W. Grove. *CAP Coastal Patrol Base Twenty-one.* Raleigh, NC: Bynum Printing Co., 1944.

Wiggins, Melanie. *Torpedoes in the Gulf: Galveston and the U-Boats, 1942–1943.* College Station: Texas A&M University Press, 1995.

Wilson, Gill Robb. *I Walked with Giants.* New York: Vantage Press, 1968.

Published and Unpublished Papers

Batten, John H. Civil Air Patrol National Historical Committee. *Civil Air Patrol Historical Monograph Number 7: History of Coastal Patrol Base No. 6, Civil Air Patrol, World War II, St. Simons Island, Georgia.* Maxwell AFB, AL: CAP National Headquarters, 1988.

Blazich, Frank A., Jr. Civil Air Patrol National History Program. *"Founding" versus "Establishment": A Perspective on Civil Air Patrol's Recognized Date of Origin.* Maxwell AFB, AL: CAP National Headquarters, 2017.

———. "'Definitely Damaged or Destroyed': Reexamining Civil Air Patrol's Wartime Claims." *Air Power History* 66, no. 1 (Spring 2019): 19–30.

———. "Earle L. Johnson: CAP's Wartime Commander." *CAP National Historical Journal* 1, no. 2 (January–March 2014): 9–15.

———. "Economics of Emergencies: North Carolina, Civil Defense, and the Cold War, 1940–1963." PhD diss., The Ohio State University, 2013.

———. "North Carolina's Flying Volunteers: The Civil Air Patrol in World War II, 1941–44." *North Carolina Historical Review* 89, no. 4 (October 2012): 399–442.

Burnham, Isaac W., II. "History of CAP Coastal Patrol No. 4." N.p., [1943?].

Civilian Air Reserve (CAR). *Organization Handbook.* Toledo, OH: Civilian Air Reserve, 1940.

Cooling, B. Franklin. "U.S. Army Support of Civil Defense: The Formative Years." *Military Affairs* 35, no. 1 (February 1971): 7–11.

Gettinger, Dan. "The Surrogate Predator Program." Center for the Study of the Drone. Bard College, 25 January 2016, https://dronecenter.bard.edu/the-surrogate-predator-program/.

Grebe, Earl C. "Toledo's C.A.R." *Popular Aviation* 25, no. 6 (December 1939): 35–36.

Green, Barbara G. "The History of Coastal Patrol Base #14, Civil Air Patrol, World War II, Panama City, Fla." January 1979.

Hodges, Jeremy K. "Using Auxiliary Forces to Accomplish Strategic Objectives." *Air & Space Power Journal* 32, no. 2 (Summer 2018): 50–59.

Hopper, Lester E. *Civil Air Patrol Historical Monograph Number 1: Duck Club.* Maxwell AFB, AL: CAP National Headquarters, Civil Air Patrol National Historical Committee, 1984.

———. *Civil Air Patrol Historical Monograph Number 2: Air Medal.* Maxwell AFB, AL: CAP National Headquarters, Civil Air Patrol National Historical Committee, 1984.

Hudson, James J. "Reed G. Landis and 'The Last Good War.'" *Arkansas Historical Quarterly* 35, no. 2 (Summer 1976): 127–41.

Hudson, Seth. Civil Air Patrol National History Program. *Civil Air Patrol Fatalities, 1941–Present*. Maxwell AFB, AL: CAP National Headquarters, 29 May 2020.

"Inexpensive Bombing for Civil Air Patrol." *Air Force: Official Service Journal of the U.S. Army Air Forces* (January 1943): 34.

Joiner, Stephen. "Cessnas Pretending to Be Predators." *Air and Space Magazine* (June 2016): https://www.airspacemag.com/flight-today/soundings-cessna-180959131/.

Jordan, Nehemiah. *U.S. Civil Defense Before 1950: The Roots of Public Law 920*. Washington, DC: Institute for Defense Analyses, May 1966.

Keil, John A. "Civil Air Patrol Coastal Patrol No. Three, Lantana, Florida." N.p., [1943?].

King, Elizabeth W. "Heroes of Wartime Science and Mercy." *National Geographic* 84, no. 6 (December 1943): 715-40.

Miller, Robert E. "The War that Never Came: Civilian Defense, Mobilization, and Morale During World War II." PhD diss., University of Cincinnati, 1991.

Morse, Louisa S. *Civil Air Patrol Uniforms and Insignia: The First Ten Years, 1941–1951*. Maxwell AFB, AL: Civil Air Patrol National Headquarters, Civil Air Patrol Historical Committee, 1982.

Nanney, Claude Y., Jr. *CAP Coastal Patrol Base No. 5, Flagler Beach, Florida, May 19, 1942 . . . August 31, 1943*. Daytona Beach, FL, 31 August 1943. http://archives.sercap.us/resources/site1/General/Coastal_Patrol/Base_05/Coastal-Patrol-Base-No-5.pdf.

Neprud, Robert E. *Civil Air Patrol in World War II*. Washington, DC: Army Times, 1946.

Reilly, Thomas. "Florida's Flying Minute Men: The Civil Air Patrol, 1941–1943." *The Florida Historical Quarterly* 76, no. 4 (Spring 1998): 417–38.

Speiser, Stuart M. "'Joe'—Submarine Hunter—1942–1943: A True Story of Grand Isle Patrol Activity." N.p., [1943?].

Weidenfeld, Gregory. "The Search for the Haggin-Farr Sub Kill." McGuire ARB, NJ: New Jersey Wing Headquarters, 1997.

"What Is a 'Barracuda Bucket?'" *The Barracuda Bucket: Official Journal of Florida Wing* 41 (May 1944), 1.

Government Publications

Air Force Instruction 10-2701. *Organization and Function of the Civil Air Patrol*, 7 August 2018.
Air Force Instruction 52-101. *Chaplain, Planning and Organizing*, 15 July 2019.
Civil Air Patrol. *2018 Report to Congress: 70 Years as U.S. Air Force Auxiliary*. Maxwell AFB, AL: CAP National Headquarters, 2018.
_____. *2019 Annual Report*. Maxwell AFB, AL: CAP National Headquarters, 2019.
_____. CAP Manual 50-1. *Introduction to Civil Air Patrol*. January 1958.
_____. *Civil Air Patrol Manual: An Introduction to Civil Air Patrol, an Auxiliary of the U.S. Air Force*. Vol. 1, Book 1. Washington, DC: GPO, 1949.
_____. *Civil Air Patrol–US Air Force Auxiliary 2016 Financial Report*. Maxwell AFB, AL: CAP National Headquarters, 2017.
_____. Office of the Staff Judge Advocate. *The History of the Legal Status and Administrative Concepts of Civil Air Patrol*. Maxwell AFB, AL: CAP National Headquarters, June 1985.
Craven, Wesley Frank, and James Lea Cate, eds. *Plans and Early Operations, January 1939 to August 1942*. Vol. 1 of *The Army Air Forces in World War II*. 1949. Reprint, Washington, DC: Office of Air Force History, Government Printing Office, 1983.
_____. *Europe: Torch to Pointblank, August 1942 to December 1943*. Vol. 2 of *The Army Air Forces in World War II*. 1949. Reprint, Washington, DC: Office of Air Force History, Government Printing Office, 1983.
Douhet, Giulio. *The Command of the Air*. Translated by Dino Ferrari. 1942. Reprint, Washington, DC: Office of Air Force History, Government Printing Office, 1983.
Finney, Robert T. *History of the Air Corps Tactical School, 1920–1940*. 1955. Reprint, Washington, DC: Center for Air Force History, 1992.
Frey, John W., and H. Chandler Ide, eds. *A History of the Petroleum Administration for War, 1941–1945*. Washington, DC: Government Printing Office, 1946.
Graham, William R., et al. *Report of the Commission to Assess the Threat to the United States from Electromagnetic Pulse (EMP) Attack,*

Vol 1: Executive Report (2004), http://www.empcommission.org/docs/empc_exec_rpt.pdf.

Gross, Charles J. *Prelude to the Total Force: The Air National Guard.* Washington, DC: Office of Air Force History, Government Printing Office, 1985.

Holley, Irving B., Jr. *Army Air Forces Historical Studies No. 44: Evolution of the Liaison-Type Airplane, 1917–1944.* Washington, DC: Headquarters Army Air Forces, AAF Historical Office, April 1946.

Huston, John W., ed. *American Airpower Comes of Age: General Henry H. "Hap" Arnold's World War II Diaries.* Vol. 1. Maxwell AFB, AL: Air University Press, 2002.

Link, Mae M. *Army Air Forces Historical Studies No. 19: Civilian Volunteer Activities in the AAF.* Washington, DC: Assistant Chief of Air Staff, Intelligence, Historical Division, October 1944.

Mauck, Elwyn A. "Civilian Defense in the United States, 1941–1945." Microfilm. Unpublished manuscript by the Historical Officer of the Office of Civilian Defense, July 1946, typed.

Maurer, Maurer. *Air Force Combat Units of World War II.* 1961. Reprint, Washington, DC: Office of Air Force History, Government Printing Office, 1983.

Meilinger, Philip S. *Bomber: The Formation and Early Years of Strategic Air Command.* Maxwell AFB, AL: Air University Press, 2012.

_____., ed. *The Paths of Heaven: The Evolution of Airpower Theory.* Maxwell AFB, AL: Air University Press, 1997.

Navy Department. Chief of Naval Operations. *German, Japanese and Italian Submarine Losses, World War II.* Washington, DC: Navy Department, OPNAV-P33-100, May 1946.

New Jersey Defense Council. *New Jersey Wing: Civil Air Defense Services.* Trenton: New Jersey Defense Council, November 1941.

Raines, Edgar F., Jr. *Eyes of the Artillery: The Origins of Modern U.S. Army Aviation in World War II.* Washington, DC: Center of Military History, 2000.

Rosenfeld, Susan, and Charles J. Gross. *Air National Guard at 60: A History.* Washington, DC: Air National Guard, Government Printing Office, 2007.

Spink, Barry L. "Distinguished Flying Cross and Air Medal Criteria in the Army Air Forces in World War II." Montgomery, AL: Air Force Research Agency, Maxwell AFB, 2015.

Strickland, Patricia. *The Putt-Putt Air Force: The Story of the Civilian Pilot Training Program and the War Training Service (1939–1944)*. Washington, DC: Department of Transportation, Federal Aviation Administration, Aviation Education Staff, 1971.
Tilley, John A. *The United States Coast Guard Auxiliary: A History, 1939–1999*. Washington, DC: GPO, 2003.
United States Air Force. Assistant Chief of Air Staff, Intelligence. Historical Division. *Historical Study No. 107–The Antisubmarine Command*. Washington, DC: US Army Air Forces, April 1945.
United States Office of Civilian Defense (OCD). *Civil Air Patrol: Organization, Purpose, Program, Enlistment*. Washington, DC: GPO, 1941.
Wilson, John R. M. *Turbulence Aloft: The Civil Aeronautics Administration Amid Wars and Rumors of Wars, 1938–1953*. Washington, DC: GPO, 1979.
Yoshpe, Harry B. *Our Missing Shield: The U.S. Civil Defense Program in Historical Perspective*. Washington, DC: Federal Emergency Management Agency, April 1981.

Newspapers and Periodicals
(specific articles and authors indicated in endnotes)

Air Force Magazine
AIRMAN Magazine
Amarillo (TX) Globe-Times
Asbury (NJ) Park Press
Baltimore Afro-American
Bangor (ME) Daily News
Boston Globe
CAP Bulletin
CAP News
Charlotte (NC) Observer
Central New Jersey Home News (New Brunswick, NJ)
Chicago Daily Tribune
Chicago Defender
Cincinnati Enquirer
Civil Air Patrol Volunteer
Courier-News (Bridgewater, NJ)
Courier-Post (Camden, NJ)
Daily Journal (Vineland, NJ)

Daily Press (Newport News, VA)
Daily Times (Salisbury, MD)
Dare County Times (Manteo, NC)
Delaware County Daily Times (Chester, PA)
Flying and Popular Aviation
Hartford (CT) Courant
Knoxville (TN) Journal
Life
Miami (FL) News
Monitor (McAllen, TX)
National Aeronautics
Nebraska State Journal (Lincoln)
New York Daily News
New York Times
News and Observer (Raleigh, NC)
News Journal (Wilmington, DE)
Palm Beach Post (West Palm Beach, FL)
Philadelphia Inquirer
Rocky Mount (NC) Telegram
Sentinel (Carlisle, PA)
State Journal (Lansing, MI)
Syracuse (NY) Post-Standard
Walt Disney's Comics and Stories
Washington Post

Interviews Conducted by Frank A. Blazich Jr.

Arn, Robert E. Westerville, OH, 31 March 2012.
Jividen, Carl E. Londonderry, OH, 19 April 2012.

Interview Conducted by David E. Gillingham

Compton, Charles E. Evanston, IL, 19 June 2015.

Interviews Conducted by Lester E. Hopper

Bridges, William Paul. Manteo, NC, 10 October 1983.
Edwards, Edmond I. Rehoboth, DE, 14 September 1990.

Eggenweiler, Marilou Crescenzo. El Paso, TX, 19 May 1984.
Fandison, William J. New Orleans, LA, 1 March 1982.
Gantt, William M. Sea Island, GA, 22 January 1986.
Grier, W. Logan. Milford, DE, 14 September 1991.
Haddaway, George E. Rehoboth, DE, 14 May 1983.
McDonald, Addis H. Little Rock, AR, 17 September 1983.
Neprud, Robert E. Las Vegas, NV, 12 August 1983.
Parkinson, Marion F. Pascagoula, MS, 9 March 1983.
Sharp, Hugh R., Jr. Wilmington, DE, 17 October 1983.
Swaim, Carl O. Manteo, NC, 10 October 1983.
Wescott, Dorothy Graham. Manteo, NC, 10 October 1983.

Interview Conducted by Philip C. Saleet

Fields, Charles W. Sr. Greensboro, NC, 8 August 2012.

Interview Conducted by Hellenmerie Walker

Myers, Frank S. Portland, OR, 18 July 1983.

Websites

Ancestry, http://www.ancestry.com
CAP National History Program, http://history.cap.gov
Civil Air Patrol, http://www.gocivilairpatrol.com
Fold3, http://www.fold3.com
New England Air Museum, https://www.neam.org/
Sun Transport Fleet Sheet, http://www.fleetsheet.com
Uboat.net, http://www.uboat.net
U-Boat Archive, http://www.uboatarchive.net
United States Air Force, http://www.af.mil

Index

I Air Support Command, 48, 51, 53, 57, 58, 74, 100n100, 118
I Bomber Command, 39, 40, 46, 57, 100, 114–17, 149
I Ground Air Support Command, 84, 87–88, 97n100, 100, 114
I Patrol Force, 114, 115
III Fighter Command, 5
1st Air Squadron, 58, 59
25th Antisubmarine Wing, 117, 149, 160n113
25th Pursuit Squadron, 24
26th Antisubmarine Wing, 117, 149, 160n113
59th Observation Group, 51, 114
65th Observation Group, 114
104th Observation Squadron, 51, 53, 56
120th Observation Squadron, 118
124th Observation Squadron, 100
128th Observation Squadron, 100

A-1-A priority, 139, 140
African Americans, 73
Air Corps Enlisted Reserve (ACER), 138, 140
Aircraft Owners and Pilots Association (AOPA), 17–18, 24–26
aircrew fatigue and rest, 134, 167
Air Force, 2–5, 11, 21, 22, 39, 41, 43, 45, 49, 53, 56, 81, 85, 86, 108, 114, 115, 117, 118, 141, 143, 145, 147, 169, 172
Air Forces Northern, 60n3. *See also* First Air Force
airframe and engine (A&E), 130, 177
Air Guard, 18, 20, 24–26
Air Medal, 111, 112, 123n56, 169
air-mindedness, 21–23
Air National Guard, 172, 173, 178
Air Vigilante, 22, 25
Andrews, Adolphus, 39, 40, 46, 55, 56, 57, 58, 67, 75, 80, 93n47, 99, 104, 142, 145, 150, 163

Army Air Forces, 1–3, 6, 8n20, 22, 25, 36n78, 38n90, 41, 43–45, 48, 49, 56, 57, 59, 62n34, 74, 76, 80, 81, 83, 86, 88, 100, 105, 106, 112, 116, 118, 129, 130, 134, 137, 138, 139–40, 144, 146–47, 149, 158n90, 167–71, 199, 200
Army Air Forces Antisubmarine Command (AAFAC), 116–18, 135, 140–43, 148, 149, 156n69, 158nn89–90
Arnold, Henry H., 22–24, 39, 41–44, 46, 48–50, 56, 80–83, 85–87, 96n94, 106, 116, 118, 138–141, 146, 150, 155n62, 163, 169, 170–71, 182n19, 199, 200, 203n11
auxiliary aviation, 13, 17–18, 170
auxiliary status, 172

Bane, Frank, 16, 17
Bard, Ralph, 85
Battle of the Atlantic, 1, 62n45, 83, 89, 165, 166, 169, 179
Beck, Thomas H., 18, 19, 21, 22, 25
belligerent (status), 68, 167, 170
Blee, Harry H., 67–69, 72, 80, 83, 105, 106, 114, 139, 168, 199, 200
bombs: AN-M30, 88, 135, 198; AN-M57, 135, 196, 198; Mk 17, 135, 198; bombing accuracy, 132, 135
bombsight, 86–87, 135, 154n38, 197, 198, 202nn6–7
Bradley, Follett, 56, 64n82, 81, 85, 86, 100
Bureau of Ordnance, 145
Bureau of the Budget, 28, 42, 43, 57, 67, 104, 126n84, 157n79
Burnham, Isaac W., 3, 67, 76–78, 80, 143, 144, 148, 159n106, 191

camouflage, 140, 141
Cape Hatteras, NC, 47, 48, 56, 80, 108
Certificate of Military Necessity, 139, 140

INDEX

Civil Aeronautics Administration (CAA), 16, 19, 22–23, 24, 25, 28, 43, 55, 86, 89, 130, 134, 143

Civil Aeronautics Board, 132, 133

Civil Air Defense Services (CADS), 19–20, 21, 22–25, 75, 205

Civil Air Patrol aircraft, 41, 46, 50, 52, 53, 55–57, 67–69, 84–88, 100, 104, 108, 112, 115, 118, 132, 135, 145, 147, 164, 168; instrumentation, 165; requirements, 28, 45, 79, 82, 86, 114, 129, 134, 139, 143, 167, 174, 197, 200

Civilian Air Reserve (CAR), 13–16, 19, 24–25

Civilian Pilot Training Program (CPTP), 18, 23, 33n56, 143

Civil Reserve Air Fleet, 179

coastal patrol bases, 51, 64n82, 71–73, 78, 80, 82, 88, 89, 92n36, 92n39, 93n43, 98n106, 100, 101, 107, 111, 113, 115, 116, 124n61, 130, 131, 133, 137, 139, 141, 142, 147, 151, 191; budgets, 28, 42, 43, 49, 57, 58, 67, 82, 83, 104, 146

locations:
    Atlantic City, NJ, 47–49, 53–56, 58, 59, 67, 69, 74, 76, 81, 117, 187, 191, 197, 198
    Bar Harbor, ME, 130, 188, 192, 197
    Beaufort, NC, 113, 130, 132, 133, 136, 141, 146–148, 188, 192, 197
    Beaumont, TX, 100, 133, 187, 191, 197
    Brownsville, TX, 5, 72, 119, 142, 191
    Charleston, SC, 132, 133, 187, 190, 191, 197
    Corpus Christi, TX, 187, 191, 197
    Falmouth, MA, 187, 192, 197
    Flagler Beach, FL, 73, 187, 191, 197
    Grand Isle, LA, 100–102, 129, 131, 187, 191, 197
    Lantana, FL, 88, 117, 187, 191, 197
    Manteo, NC, 4, 47, 108, 116, 130, 136, 148–150, 187, 191, 197
    Miami, FL, 50, 84, 85, 117, 131, 136, 187, 191, 197
    Panama City, FL, 111, 129, 187, 191, 197
    Parksley, VA, 67, 76–78, 80, 100, 110, 111, 117, 143, 148, 151, 187, 191, 197
    Pascagoula, MS, 71, 100, 187, 191, 197
    Portland, ME, 118, 137, 188, 192, 197
    Rehoboth, DE, 47–49, 51–54, 58, 69, 74, 108–111, 132, 148, 187, 191, 197, 199
    Sarasota, FL, 187, 191, 197
    St. Simons Island, GA, 131, 187, 191, 197
    Suffolk, Riverhead, NY, 107, 108, 131, 137

coastal patrol experiment, 43, 48–519, 81–82, 168

Connolly, Donald H., 19, 42, 56

convoy, 2, 40, 75, 83, 93n47, 104, 115, 117, 141, 149, 165, 166

Council of National Defense, 15–17

Cremer, Peter-Erich, 165; comments on CAP, 165. *See also* U-boat operations

Crenshaw, Russell S., 99, 100

Cross, Henry T., 13, 14, 17, 24, 109–12, 123n45

Curry, John F., 26–29, 35n78, 36n81, 36n87, 40–49, 56, 68–69, 72–73, 75, 86, 140, 166, 168

Dawson, Frank E., 130, 192

Department of Commerce, 16, 22, 42, 45, 67

depth charge, 2, 84, 87–89, 97n100, 117, 135, 199–200

Desmarais, John W., Jr., 175

deterrence, 104, 166

Doenitz, Karl, 39, 45, 46, 83, 99, 104, 108, 141, 165, 166, 168

Douhet, Giulio, 21–22

Drum, Hugh A., 39, 42, 46, 49, 57, 65n89

Duck Club, 137

Eastern Defense Command, 39, 42, 149

INDEX | 237

Eastern Sea Frontier, 39, 45, 60n14, 75, 99, 142–43, 145–46, 150, 197, 198
Edison, Charles, 20, 25
Edmondson, Shelley S., 109–11
Edwards, Edmond I., 109–12, 123n45, 123n56, 169
Edwards, Richard S., 56, 63n62
Eisenhower, Dwight D., 87
electromagnetic pulse (EMP), 179, 185n48
executive orders, 19, 37n91, 146

forward air controller (FAC), 177
Fairchild Model 24, 51, 53, 70, 109, 131–33, 153n19, 167
Farr, Wynant G., 54, 64n70, 191, 198, 203n11
Federal Emergency Management Agency (FEMA), 3, 173, 179
First Air Force (Air Force North, 1AF [AFNORTH]), 3, 39, 49, 53, 56, 60n3, 81, 85, 86, 114, 115, 172, 174, 180
Florida Defense Force, 58

Gannett, Guy, 17, 19, 21, 22, 25
Gulf Sea Frontier, 99, 100, 117, 202n5, 204n31
Gulf Task Force, 100, 143

Haggin, John B., 198, 203n11
Hartranft, John B., 24
Hinckley, Robert H., 23, 28
Hoiriis, Holger, 54–55, 191
Horne, Frederick J., 85
Hoyt, Kendall K., 21, 24, 200

inspectors, 117, 132, 133
intelligence, surveillance, and reconnaissance (ISR), 177

Johnson, Earle L., 47, 72, 75–76, 80, 85, 102–3, 106, 126n84, 129–30, 138–41, 148, 164–65, 167–68, 170, 188, 199, 201
Joint Army–Navy Assessment Committee, 198–99, 201

Kauffman, James L., 100, 104
King, Archibald, 68–69

King, Ernest J., 40, 55–57, 59, 75, 93n47, 100, 103, 104, 116, 143, 146, 148, 150, 157n79, 163
King, Henry, 37n92, 41
Knight, Milton, 13–17, 24
Knox, Frank, 25
Kriegsmarine (Nazi Germany's navy), 201
Kriegstagebüch (KTB, war diary), 54, 198, 199, 202n5, 203n14
Kuter, Laurence S., 83, 87

LaGuardia, Fiorello, 11, 18, 19, 21–25, 27, 29, 35n78, 35n81, 41, 42, 44, 104, 107
Landis, James M., 42, 75, 82, 112, 144–46, 157n84
Landis, Reed G., 24–26, 29, 34n68, 35n78, 35n81, 42–43, 44–45, 61n34, 86, 166
Larson, Westside T., 141
liaison patrol, 5, 118–19, 138, 147, 169
lifesaving equipment, 113, 136, 167; life rafts, 52, 113, 136; life vests, 51, 54, 78, 136, 137, 155n45
Lovett, Robert A., 50, 55, 63nn58–59, 80–81, 144–46

manufacturers, 70, 134, 140, 167
Marshall, George C., 48, 49, 57, 84–86, 103, 116, 148, 169
Mason, William D., 46, 47, 50, 53, 63n62
Mayfair, 113
McNair, Leslie J., 49
mechanics, 11, 14, 19, 20, 22, 26, 51, 70, 71, 79, 101, 130–32, 138, 177, 197
milchkühe ("milk cow," supply boats), 99. *See also* U–boat operations
military courier, 5
Missing-Aircraft Search Service, 6, 8nn20–21
Moody, Warren E., 107, 132
MQ-9 Reaper, 173, 178
Murrow, Edward R., 16
MX-15, 173, 175

National Aeronautic Association (NAA), 17, 21, 50

National Defense Advisory Commission (NDAC), 15–17
navigation, 14, 20, 52, 61n14, 68, 69, 73, 117
Navy, 1–4, 6, 19, 25, 27, 39–41, 45, 46, 50, 51, 54–58, 74, 75, 80–85, 99, 103, 104, 112, 114, 116, 118, 130, 135, 140–49, 163, 166, 177, 180, 197–201, 202n5, 204n28, 204n30

Office of Civilian Defense (OCD), 1, 2, 4, 19, 22–29, 35n81, 37n90, 37n92, 42, 44–45, 68, 72, 73, 82, 83, 105, 107, 112, 144, 146, 166, 169, 182n18, 199, 201; Aviation Planning Staff, 29, 37n92, 41
Office of Emergency Management (OEM), 15

Panam, 146, 147
patrolling, 13, 40, 41, 48, 49, 55, 81, 118, 126n84, 150, 166, 198, 199, 202n5
Paukenschlag ("Drumbeat"), 39, 45. *See also* U-boat operations
per diem, 49, 57, 70, 89
Petroleum Industry War Council, 50
Pew, J. Howard, 46
Public Law 79-476, 171, 172
Public Law 80-557, 172

radio, 14, 16, 18, 20, 21, 26, 28, 41, 51–53, 68, 71, 72, 78, 79, 81, 89, 101, 109–111, 114, 116, 130, 131, 174, 188; contact reports, 53, 56–57, 84; procedures, 52–53, 71
reduction in flying hours, 143
Reichsluftfahrtministerium (Nazi Germany Ministry of Aviation), 12
remotely piloted aircraft (RPA), 173, 177, 178
Roosevelt, Franklin D., 15–19, 23, 57, 68, 111, 112, 123n56, 146

Saville, Gordon P., 43–44, 46
Scarecrow Patrol, 55–57
Selective Service, 28, 138
Sharp, Hugh R., Jr., 109–112, 169, 191
Shelfus, Charles E., 109, 111, 189
Sikorsky S-39B, 110–11, 123n48

small unmanned aerial system (sUAS), 173–75, 177
Smith, Everett M., 132
Spaatz, Carl A., 56, 63n62, 170
Stephan, Audley H. F., 17
Stimson, Henry L., 22, 25, 138, 146
Stinson Voyager 10A, 70, 76, 88, 134, 141, 165
Stratemeyer, George E., 111
submarine, 1, 12, 41, 43, 50, 53, 67, 74, 75, 83–85, 99, 103, 104, 107, 115–17, 131, 135, 141–42, 145, 149–50, 163, 164–66, 180, 197–202
Sumners, Hatton W., 1, 171
Sun Oil Company, 46–49, 50, 62n45, 77

tankers, 46, 50–51, 53, 57, 81, 84, 99, 101, 103, 168
task forces, 124n61. *See also* coastal patrol bases
Taylor, Irving H., 41, 140
Tenth Fleet, 198, 199, 201, 202n5, 203n8
total force, 172, 178, 180
tow target training, 5, 9n21, 25, 160n112, 169
training, 2, 3, 5, 14, 18–20, 23, 25, 26, 29, 39, 42, 59, 68, 74, 75, 79, 83, 88, 100, 114, 135, 141, 143, 166, 170–173
Truman, Harry S., 146, 169, 171, 172
Twining, Nathan F., 49, 57

U-boat operations, 99, 168. doctrine to aircraft, 67, 104; Heino von Heimburg, 165; *U-85*, 46, 81, 204n30; *U-109*, 83–85; *U-129*, 146; *U-333*, 83, 84, 165; *U-458*, 108; *U-564*, 83–85, 87; *U-754*, 108, 204n30; vulnerability, 67. *See also* Cremer, Peter-Erich; *milchkühe*; *Paukenschlag*

Ulio, James A., 82, 118
uniforms, 14, 18, 29, 54, 70, 105–8

Vermilya, Wright "Ike," Jr., 47, 58–59, 65n95, 80, 96n86, 96n94, 191
Virdin, Carl L., 109, 111
Volney, Joseph, 11
Vorys, John M., 1, 7n2, 63n59

War Department, 1, 2, 4, 22–23, 25, 26, 44, 45, 48, 56, 68–70, 76, 82–83, 97n100, 105–7, 116, 130, 138, 144–46, 150, 168–69, 182n18, 199–200

war diary, 39, 54, 75, 197, 198, 200

Weather Bureau, 115

Wilson, Gill Robb, 11–13, 15–17, 19–26, 29, 30, 54, 59, 75, 104, 144, 191

women, 72–73, 100, 105, 149, 175